ENERGY VICTORY

ADVANCE PRAISE FOR
ENERGY VICTORY

"Robert Zubrin's *Energy Victory* highlights fuel choice in the transportation sector as a real-world option and imperative to reducing America's dependence on the petrostates that fund radical Islam. It is a compelling read."

Anne Korin
Executive director of Set America Free Coalition
and editor of *Energy Security*

"In forcefully making the case for reducing US dependence on oil, Zubrin argues provocatively for a biofuel-based approach, suggesting benefits for international development that should command the attention of advocates, academics, and policymakers."

Louis Putterman
Professor of Economics
Brown University

WINNING THE WAR ON TERROR
BY BREAKING FREE OF OIL

ENERGY VICTORY

ROBERT ZUBRIN

Published 2009 by Prometheus Books

Inquiries should be addressed to
Prometheus Books
59 John Glenn Drive
Amherst, New York 14228–2119
VOICE: 716–691–0133, ext. 210
FAX: 716–691–0137
WWW.PROMETHEUSBOOKS.COM

13 12 11 10 09   5 4 3 2 1

Library of Congress Cataloging-in-Publication Data

Zubrin, Robert.
Energy victory : winning the war on terror by breaking free of oil / Robert Zubrin.
p. cm.
Includes bibliographical references and index.
ISBN 978–1–59102–707–2
1. Alcohol as fuel—United States. 2. Petroleum as fuel—United States. 3. Energy policy—United States. 4. Federal aid to terrorism prevention—United States. I. Title.

TP358.Z83 2007
333.790973—dc22

2007027122

Printed in the United States of America on acid-free paper

To the people of the United States of America—still the last, best hope of mankind—whose lives, fortunes, and freedoms are being gravely endangered by their leaders' continued failure to act responsibly for energy security.

Thus the men of democratic times require to be free in order to procure more readily those physical enjoyments for which they are always longing. It sometimes happens, however, that the excessive taste they conceive for these same enjoyments makes them surrender to the first master who appears. . . . These people think they are following the principle of self-interest, but the idea they entertain of that principle is a very crude one; and the better to look after what they call their own business, they neglect their chief business, which is to remain their own masters.

—Alexis de Tocqueville, *Democracy in America*, 1840

Only the wise can be free.

—Chrysippus, Greek philosopher, third century BCE

CONTENTS

ACKNOWLEDGMENTS

I wish to acknowledge the help of my colleagues at Pioneer Astronautics—Mark Berggren, Nick Jameson, Dan Harber, James Kilgore, Heather Rose, Tony Muscatello, Cherie Wilson, Stacey Carrera, Douwe Bruinsma, and Emily Bostwick-White—for many useful technical discussions that have helped to sharpen the contents of this book. Thanks also to Professor Louis Putterman of Brown University for providing advice on economic aspects of the matter, and to the very knowledgeable international trade writer Laura Chasen Cohen for providing insights into how biofuel policy could be used to provide maximum benefit for international trade and third world development. Thanks also to the intrepid energy independence lobbyists Gal Luft and Anne Korin and to *New Atlantis* editor Adam Keiper for many useful discussions about the political side of the battle. Special thanks are due to Karl Zinsmeister, former editor of the *American Enterprise*, who had the courage and vision to bring my first article on this subject into print, and thus start me on the road that led to this book. Thanks are also due to Gary Yowell and Tom McDonald of the California Energy Commission, for filling me in on the history of the CEC's program that led to the creation of the first flex-fuel cars. Special thanks are also due to my hardworking agent, Laurie Fox, who sold the book; to Prometheus Books editor in chief Steven L. Mitchell, who guided me to its successful completion; to Prometheus's excellent production manager, Christine Kramer, and copy editor Meghann French for their fine work; and to Prometheus's graphic designer Nicole Lecht, who designed and created its terrific cover.

Most of all I want to thank the members of my family—especially my daughter, Rachel, and my wife, Maggie—who had to put up with a stressed-out, nearly absent husband and father during this work's gestation, but whose loving support through it all made this book possible.

PREFACE TO THE PAPERBACK EDITION

Much has happened since the first hardcover edition of *Energy Victory* appeared in print in November 2007. At the time of publication, the price of petroleum stood at a near record of $70 per barrel. During subsequent months, as the OPEC cartel continued its policy of constricting production, the oil price soared to over $140 per barrel, after which the world economy collapsed.

The latter two events were not unrelated. The economic crash has been widely ascribed to the undermining of financial institutions by speculative mortgage-backed securities, made worthless by the fall of the housing market. Yet, while much ink has been spilt attempting to ascribe blame to various politicians for the poor regulation that allowed such vulnerable credit instruments to be created, we need to ask the fundamental question: *Why did the* global *housing market collapse?* If you want the answer, just follow the money. It went to pay for oil.

Consider: Over the course of Fiscal Year 2008, beginning October 2007 and running through September 2008, with OPEC-rigged oil prices averaging near $110 per barrel, Americans paid $900 billion for their oil supply, and the world as a whole paid $3.6 trillion. These petroleum costs were up a factor of ten from what they were in FY 1999, and they represent a huge, highly regressive tax on the world economy. For Americans, the $900 billion oil levy (up from $80 billion in 1999) was equivalent to a 33 percent increase in income taxes across the board—with 60 percent of the sum being paid over in tribute to foreign governments.

Averaged over the US population of 300 million people, the $900 billion OPEC tax levied a tribute amounting to $3,000 per head—for every man, woman, and child in the country, or $12,000 for a family of four. The average American worker makes about $45,000 per year, or $35,000 after taxes paid to Uncle Sam. In 1999 such a worker supporting a family of four had to pay 3 percent of his disposable income for oil. In FY 2008 Uncle Saud and Uncle Hugo hit him for over a third of his take-home pay. Is it any wonder that such people stopped buying houses and cars? Such a massive drain of cash from the pockets of consumers must perforce collapse the real estate market—as well as that for many other kinds of consumer goods.

To see how this tax can destroy real estate values, it is only necessary to compare expenditures. In FY 2003 Americans paid $268 billion for new homes and $197 billion for oil. In FY 2008 we paid for new homes at an annual rate of $134 billion and $900 billion for oil. So the increase in our oil expenditures was more than *five times* as great as the fall in our spending for new homes.

If we consider existing home sales, the pattern remains the same. An existing home held for the typical sixteen years can, on average, be expected to sell for about 40 percent over its previous purchase price. Taking this into account along with the new home sales, then total net American investment in housing stock went from around $780 billion in FY 2003 to an annual rate of $620 billion in FY 2008, a $160 billion decline that is dwarfed by the $700 billion rise in annual oil payouts over the same period.

The graph below shows a comparison of American oil expenditures with net investment in housing stock.

It can be seen that in FY 2003 we spent 25 percent as much on oil

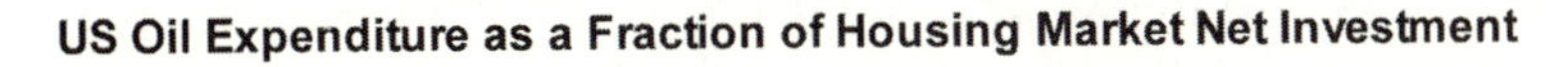

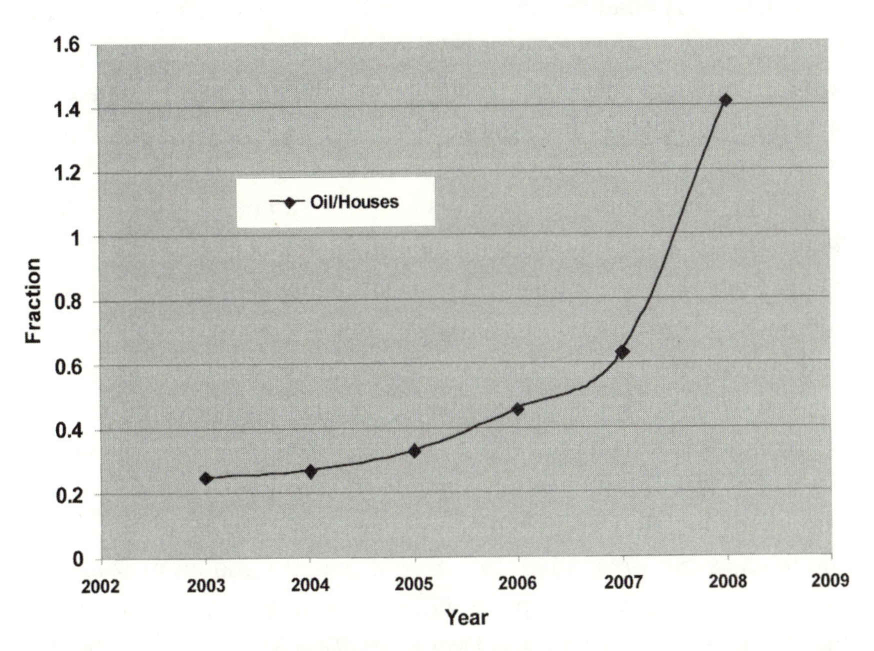

Figure 1. Comparison of American spending on oil with net investment in housing stock for fiscal years 2003–2008. Data from US DOE Energy Information Administration, www.eia.doe.gov (accessed November 15, 2008), and National Association of Realtors, http://www.realtor.org/research/research/ehspage (accessed November 15, 2008).

as we did on houses. In FY 2008, we spent over 140 percent. This is why our housing market—and that of every other industrial nation—has collapsed. It is also why new auto sales—down around $50 billion per year in the United States since 2003—have collapsed as well.

So, as a result of this massive new tax on our economy—by far the largest in American history—the United States is being driven into a recession. But for poor countries, which can afford it even less, the effect of the brutal OPEC global taxation program is much worse. It is one thing to pay $110 per barrel when you live in a country where the average person makes $45,000 per year. It is quite another if you are an African or Haitian making $1,000 per year. The oil cartel is already starving many of these people by driving up transport, growing, and fishing costs as well as by using its petrodollar funds to engage in speculation, which in the course of last year drove farm commodity prices up by a factor of two.

During the summer of 2008, as oil prices topped $140 per barrel, there was much public outcry for an energy policy that would free the nation from OPEC's deadly grip. But subsequently, as the global economic catastrophe caused oil prices to collapse as well, attention has drifted from this priority. This is a serious mistake.

The point needs to be underlined. Yes, it is true that oil prices have finally fallen to moderate levels of $40 per barrel (still more than triple those of 1999). But this is only because the economy has fallen apart. The enemy has stopped punching us because we are flat on the mat—and his fists can't reach that low. But he is still standing over us, and as soon as we try to get up, he will slam us right back down again. Indeed, during November and December 2008, even as the leaders of the industrial nations scrambled desperately to implement stimulus programs to save their economies, OPEC's conspirators held meetings in which they agreed to take 4.5 million barrels of oil per day off the world market for the explicit purpose of forcing prices back to economic strangulation levels of over $70 per barrel as quickly as possible. If the economy should begin to recover, that chokehold is sure to send prices soaring again.

OPEC has made their future program clear; it will consist of repeated cyclic binges of vicious looting followed by recession, with each raid transferring further large increments of wealth and power to the cartel, and the reactionary Islamists and other totalitarians that stand behind it. During FY 2008 alone, OPEC's net export profits reached $1.5 trillion, an amount equal to roughly 10 percent of the October 2008 valuation of the entire US Fortune 500. Much of these profits have been placed in Sovereign Wealth Funds, whose explicit purpose is to enable the cartel's members to take over corporations in the United States and Europe. With stock prices falling, they will be able to do so at bargain prices. They are also buying up US government debt, which recession-driven deficit spending has sent completely out of control. And with each takeover, bailout, or investment operation—or generous donation to a university or political think tank—the power and influence of the cartel's money within the American and the European political systems is growing, a trend that will make effective government action to counter OPEC increasingly difficult to achieve.

Since the time of the 1973 oil embargo, when OPEC first demonstrated its malevolence, America's leaders have repeatedly declared their commitment to achieve energy security but have accomplished nothing. There is not that much time left. We do not have another three decades to waste. Each looting cycle makes the enemy stronger and ever better positioned to confound our efforts to escape.

The stakes are very high. They involve not just the future of our prosperity, but of our freedom. We need to break the power of this merciless cartel, and soon.

It can be done. Though the hour is late, we can still achieve victory. This book will explain how.

Robert Zubrin
February 11, 2009

PREFACE

America is losing the war on terror.

For the past thirty-five years, we have allowed the enemy's power to grow, and as a result, a cult that was once an anachronistic curiosity has now become a worldwide menace.

Saudi Arabia is the primary global financier of the Islamist terror cult. In 1972, Saudi foreign exchange earnings were $2.7 billion. In 2006 they topped $200 billion. Over the same period, the United States' dependency upon foreign oil grew from 30 percent to 60 percent and our annual oil import bill grew from less than $4 billion to more than $260 billion.

As a result of our failure to enact a competent energy policy, our country is being looted, and the enemy's power has been fabulously multiplied. We are financing a war against ourselves. And with the rapid industrialization of China and India increasing global demand for oil, prices are set to soar even further. Unless action is taken, things are about to get much worse.

I am an engineer. In engineering, if you want to create a successful device, you need a plan. You can't build a bridge, an airplane, or a nuclear reactor by engaging in random acts that *might* be useful toward that goal. The same is true in war. If you want to win, you need to think the matter through, develop a strategy that can actually lead to victory, and then deliberately take the steps required to implement it.

Unfortunately, we have not approached the battle for energy security in this manner. Instead, we have engaged in a wild assortment of projects with no coherent strategy behind them. Some, such as the Bush administration's hydrogen hoax, have been simply farcical. Others, such as the conservation panacea of the administration's liberal opponents or the ethanol program in its current form, actually offer some results, but not remotely enough to turn the tide. Indeed, even if one credits as real the president's grand announcement in his January 2007 State of the Union address of a goal of deriving 20 percent of all US gasoline from ethanol by 2017, the goal itself is insufficient: By 2017, US automobile fuel consumption will increase 30 percent, and the price of oil could easily double or triple. So in fact, far from being visionary, the 20 percent ethanol fuel goal by 2017 is a strategy for defeat—defeat with slightly reduced losses, perhaps, but defeat nevertheless.

America does not need a strategy for softened defeat. We need a plan for victory. In this book I will lay one out. I will show you how we can completely break the oil cartel, and with it the primary financial engine promoting the growth of the international terrorist movement. Further, I will show you how we can redirect huge flows of capital that are now going to the Saudis, the Iranians, and other OPEC bandits to farmers in our own country and around the globe. I will show you how, by doing so, we can also counter the problems of pollution and global warming in a way that is fully compatible with an open-ended future of human progress, liberty, and economic growth.

The key step to make this happen is for Congress to pass a law mandating that all new cars sold in United States be flex-fueled. Such a law would make flex-fuel the *international* standard, putting hundreds of millions of cars capable of running on alcohol fuels on the road worldwide within very few years. This step alone would break the oil cartel's vertical monopoly of the world's fuel supply, forcing them to compete with methanol and ethanol produced by farmers around the globe and thereby containing their plans for future unconstrained price hikes. By taking further steps outlined in this book, we could marginalize them completely.

Instead of financing terrorism, our energy dollars could be used to fund world development. Instead of selling blocks of our media to Saudi princes, we could be selling tractors to Africa. Instead of paying for death, we could be helping to spread life. Instead of buying arms for our enemies and chains for ourselves, we could be building a world of prosperity and freedom.

That's how we win the war on terror.

As Thomas Paine wrote to our countrymen during an earlier time of crisis, "We have it in our power to begin the world anew." And so we still do. The choice is ours.

Robert Zubrin
October 1, 2007

NOTE ON UNITS OF MEASUREMENT

In discussing energy, one unavoidably encounters a variety of units of measurement. Some are metric, some are English, some are archaic, and some are simply unique to the energy business. The table below is provided to help you sort them out.

LENGTH AND AREA

1 meter = 3.28 feet
1 kilometer = 1,000 meters = 3,280 feet = 0.62 miles
1 square meter = 10.76 square feet
1 hectare = 10,000 square meters = 2.46 acres

VOLUME

1 cubic meter = 1,000 liters = 35.38 cubic feet = 6.29 barrels = 264.15 gallons
1 barrel = 42 gallons = 159 liters

1 cubic foot = 7.47 gallons = 28.27 liters
1 gallon = 3.785 liters
1 liter = 1.057 quarts

MASS

1 kilogram (kg) = 2.2 pounds (lbs)
1 metric ton = 1 tonne = 1,000 kg = 2,200 lbs = 1.1 English tons
1 tonne of oil = 7.33 barrels of oil
1 barrel of oil = 136 kg = 300 lbs of oil
1,000 cubic feet of natural gas = 20.6 kg = 45.3 lbs of natural gas

ENERGY

1 kilocalorie (kcal) = 4.19 kilojoules (kJ) = 1.164 watt-hours = 3.968 BTU

1 kJ = 0.948 BTU

1 kilowatt (kW) = 1,000 watts = 1 kJ/s = 1.34 horsepower

1 megawatt (MW) = 1,000 kW = 1,000,000 watts = 1,340 horsepower

1 terrawatt (TW) = 1,000 gigawatts (GW) = 1,000,000 megawatts (MW)

1 kW-hour = 3,600 kJ = 3,409 BTU

1 TW-year = 3.15×10^{16} kJ = 31.5 exajoules = 29.86 quadrillion BTU

TEMPERATURE

1°C = 1°K = 1.8°F
1°F = 0.555°C
1 keV = 11,000,000°C = 19,800,000°F

MIXED UNITS

1 million barrels of oil/day = 49.8 million tonnes/year
1 million tonnes of oil/year = 20,080 barrels of oil per day
1 gallon/acre = 9.3 liters/hectare
1 liter/hectare = 0.107 gallon/acre

CHAPTER 1

ENERGY INDEPENDENCE WITHIN A DECADE

President George W. Bush has said that Iraq is "the central front in the War on Terrorism." He is wrong. The central, decisive front is America's fight for energy independence.

The world economy is currently running on a resource that is controlled by our enemies. This threatens to leave us prostrate. It must change—and the good news is that it *can* change, quickly.

In this book I will lay out the plan that will allow America and the West to emerge victorious. But before explaining the solution, let us review the problem.

Using portions of the hundreds of billions of petrodollars they are annually draining from our economy, the Saudis have set up more than twenty thousand radical madrassas around the world to indoctrinate young boys with the idea that the way to paradise is to murder Christians, Jews, Buddhists, Taoists, and Hindus.[1] The graduates of these academies killed three thousand American civilians on September 11, 2001, and have continued to kill large numbers of Americans in uniform in Iraq. Arab oil revenues are underwriting news media outlets that propagandize hatefully against the United States and the West, supporting training centers for terrorists, paying bounties to the families of suicide bombers, and funding the purchase of weapons and explosives. We are subsidizing a war against ourselves.

And we have not yet reached the culmination of the process. Iran and other states are now using petroleum lucre to underwrite the development of nuclear weapons and insulate themselves from the economic sanctions that could result. Once produced, these nuclear weapons could be used directly or be made available to terrorists to destroy US, European, Russian, or Israeli cities. This is one of the gravest threats of the next generation—and, again, we are paying for it ourselves with oil revenue.

Our responses to these provocations have been utterly muted and hapless. Why? Because any forceful action on our part against nations like Iran and Saudi Arabia could result in the disruption of the oil supplies upon which the world economy is completely dependent. We can't stand up to our enemies because we rely upon them for the fuel that is our own lifeblood.

The situation is even worse below the surface. In addition to financing terror, indirectly and directly, oil exporters are using their wealth to corrupt our political system. Important Washington, DC, law firms and lobbying organizations have been put on the payroll of Arab nations, and Saudi investors have bought enormous blocks of shares in organizations like AOL-Time Warner and News Corporation. These purchases may soon place them in position to substantially influence domestic public opinion and the governmental decision-making process, thereby enabling them to prevent either retaliation or the emergence of any effective energy policy that might break their hold on our economy.

All this, however, is mere prologue. China and India are rapidly industrializing, and within a decade or two the number of cars in the world will double or triple. If the world remains on the oil standard, the income streams of many noxious oil exporters will soar. We will be impoverished to the same degree they are enriched. The vast sums transferred will finance not only global jihad and dangerous weapons development in the Middle East, but also increased potential for manipulation of the US and Western economies. At currently projected rates of consumption, by the year 2020 more than 83 percent of the world's remaining petroleum reserves will be in the Middle East,[2] controlled by people whose religion obligates them to subjugate us.

In light of these realities, current US energy policy is a scandal. There is no reason the United States should remain helpless, allowing itself to be looted by people who are using the proceeds to undermine us. Victory in the fight for energy independence is possible—and in fact the means by which it can be achieved are now apparent. Yet victory is not being pursued. To see how insane our national energy policies have been, let's review recent failures. Then I'll describe a starkly better alternative.

THE HYDROGEN HOAX

The energy panacea of the moment is a concept called the "hydrogen economy." Theorists propose to transition US energy usage to hydrogen—a common element that, when combined with oxygen, releases energy with only water as a waste product. With hydrogen, it is claimed, we can achieve not only energy independence but also an end to pollution and global warming at the same time. As we shall discuss at greater length in chapter 6, this concept is entirely false.

Hydrogen is not a source of energy. In order to be obtained, it must be *made*—either through the use of electrolysis to split water or through the chemical breakdown of petroleum, natural gas, or coal. Either process necessarily consumes more energy than will be released by the hydrogen it produces.

When hydrogen is made by electrolysis, the process yields 85 units of hydrogen energy for every 100 units of electrical energy used to break down the water. That is 85 percent efficiency. If the hydrogen is then used in a fuel cell in an electric car, only about 55 percent of its energy value will be used; the rest is wasted as heat and so forth. The net result of these two processes: the amount of useable energy yielded by the hydrogen will be only about 47 percent as much as went into producing it in the first place. And if the hydrogen is burned in an internal combustion engine to avoid the high production costs of fuel cells, the net efficiency of this vehicle will be closer to 25 percent.

Hydrogen produced from hydrocarbons instead of water also throws away 30 to 50 percent of the total energy in the feedstock. That method actually *increases* the nation's need for fossil fuels, and thus greenhouse gas emissions increase as well. While hydrogen could also be produced by nuclear, hydroelectric, solar, or wind power, the process would continue to be dragged down by the fundamental inefficiency of hydrogen production. Such power supplies could always do more to reduce fossil fuel requirements simply by sending their electric power directly to the public grid.

The bottom line is that hydrogen is not a source of energy. It is a *carrier* of energy. And one of the least practical carriers of energy we know of.

Consider: A standard molecular weight (or mole) of hydrogen gas, when reacted with oxygen, yields 66 watt-hours of energy. Meanwhile a mole of methane (the primary component of natural gas) produces 218 watt-hours of energy. An equal number of moles of both can be stored in a tank of equal size and strength. Thus, a car that runs on compressed methane will be able to store more than three times the energy, and travel three times as far, as the same car running on hydrogen. In addition, the methane would be cheaper.

In short, from the point of view of production, distribution, environmental impact, and utility of use, the hydrogen economy makes no sense whatsoever. Its fundamental premise is at variance with the most basic laws of physics. The people who have foisted this hoax on the American political class are charlatans, and they have done the nation an immense disservice.

THE FAILURE OF CONSERVATION

The Democratic opponents of the Bush administration's hydrogen policy, while oblivious to the program's overall impossibility, have criticized it as being incapable of producing useful near-term results. In this they are certainly correct; however, as we shall see in chapter 4, their alternative policy—that of decreasing energy dependence through conservation—is inadequate and will fail strategically.

To see this, it is simply necessary to run the numbers. About 17 million cars are sold each year in the United States, or roughly 10 percent of the total in active use. If *every* consumer were to buy a hybrid car offering a 30 percent fuel saving over their existing car, and *none* of these people chose to drive more because they now had a car offering better mileage, and there were *no* expansion in the US vehicle fleet, such an innovation would result in a reduction in gasoline use of 3 percent per year. However, gasoline prices have been, and will continue, rising at a rate much higher than 3 percent, so even under this rosy scenario, the rate of looting of the US economy through fuel charges would continue to increase.

In fact, conservation offers no prospect of being even this effective. Most industry analysts predict a hybrid share of not 100 percent, but less than 1 percent of the market. But assuming that we grant such vehicles a wildly optimistic 5 percent market share, and assume another 10 percent of the market were taken by people who chose to buy ordinary economy cars offering 15 percent better mileage than their current vehicles, the net reduction of average fuel use per car would be only 0.3 percent per year. This is much smaller than the annual growth of the number of cars on the road, so under any realistic conservation scenario, total gasoline consumption would continue to rise, with rapidly increasing fuel prices radically compounding the problem.

As a method of achieving energy independence, conservation through gasoline efficiency is a losing strategy. It is like trying to use buckets to bail out a ship whose bottom has been ripped open. Or perhaps, to illustrate the issue more graphically, it is like trying to survive in a gas chamber by holding your breath. We need to break out of the gas chamber.

NUCLEAR, HYDRO, WIND, GEOTHERMAL, AND SOLAR POWER

Nuclear, hydroelectric, wind, geothermal, and solar power have all been offered as solutions to the energy problem. These all have various issues associated with them. However, the bottom line is that discussion of these technologies misses the point. It is true that, to the extent they can be done economically, these technologies can reduce greenhouse gas emissions by replacing coal or natural gas for electricity generation. But the central issue of energy independence is not electricity. The United States has plenty of coal, and if necessary it could generate all of its electric power in this way.

No, the key issue in energy independence concerns the availability of *liquid fuels* to power cars, trucks, trains, ships, and airplanes. These systems are not merely conveniences that have become dear to our way of life; they are the sinews of the economy and the fundamental instruments of military strength.

During World War II, when the fuel supplies of the Axis nations collapsed, so did their war efforts. A modern war cannot be run—a modern economy cannot be sustained—without liquid fuels.

There is no prospect whatsoever of the large-scale economic generation of liquid fuels from nuclear, hydroelectric, wind, geothermal, or solar power sources in the near future. Thus the discussion of these technologies is largely irrelevant to the immediate strategic problem we face. In the long run, however, when combined with the switch to alcohol fuels, they can play a key role in enabling an open human future of worldwide economic development and rising living standards, without the threat of global warming. This will be discussed in chapters 10 and 11.

THE ALCOHOL SOLUTION

To liberate ourselves from the threat of foreign economic domination, to destroy the economic power of the terrorists' financiers, and to give ourselves the free hand necessary to deal as forcefully as required with such people, we must devalue their resources and increase the value of our own. We can do this by *taking the world off the petroleum standard and putting it on an alcohol standard.*

This may sound like a huge and impossible task, but with gasoline prices well over $2 per gallon, the means to accomplish it are now at hand. Congress could make an enormous step toward American energy independence within a decade or so if it would simply pass a law stating that all new cars sold in the United States must be flexible-fuel vehicles (FFVs) capable of burning any combination of gasoline and alcohol. The alcohol so employed could be either methanol or ethanol.

The largest producers of both ethanol and methanol are all in the Western Hemisphere, with the United States having by far the greatest production potential for both. Ethanol is made from agricultural products. Methanol can also be made from *any kind* of biomass, as well as from natural gas or coal. American coal reserves alone are sufficient to power every car in the country on methanol for more than 250 years.[3]

Ethanol can currently be produced for about $1.50 per gallon. At this writing (August 2007) methanol is being sold, without any subsidy, for $0.93 per gallon. With gasoline having roughly doubled in price in the past three years and little likelihood of a substantial price retreat in the future, high alcohol-to-gasoline fuel mixtures are suddenly practical. As discussed in chapter 6, cars capable of burning such fuel are no futuristic dream. This year, Detroit will offer some two dozen models of standard cars with a flex-fuel option available for purchase. The engineering difference is in one sensor and a computer chip that controls the fuel-air mixture, and the employment of a corrosion-resistant fuel system. The difference in price from standard units ranges from zero to $500, with $100 being typical.

Flex-fuel cars offer consumers little advantage right now, because the high-alcohol fuels they could employ are not generally available for purchase. This is because there are so few such vehicles that it doesn't pay gas station owners to dedicate a pump to cater to them. Were flex-fuel cars made the standard, however, the fuel would quickly be made available everywhere.

If all cars sold in the United States had to be flexible-fueled, foreign manufacturers would also mass-produce such units, creating a large market in Europe and Asia as well as the United States for methanol and ethanol—much of which could be produced in America. Instead of being the world's largest fuel importer, the United States could become the world's largest fuel exporter. A large portion of the money now going to the Middle East would instead go to the United States and Canada, with much of the rest going to Brazil and other tropical agricultural nations. This would reverse our trade deficit, improve conditions in the third world, and cause a global shift in world economic power in favor of the West.

By promoting agriculture, flexible-fueled vehicles also act as global cooling agents. Plants draw carbon dioxide (CO_2) out of the atmosphere. They increase water evaporation, and the water vapor thus produced transports heat from the earth's surface to the upper atmosphere, where most of it is released into space.

The use of alcohol also reduces air pollution. In fact, environmental advantages were the motivation for the initial development of the first flexible-fueled vehicles in California in the 1980s. During the era of $1.50 per gallon gasoline, methanol pleased ecological activists, but it was economically disadvantageous to consumers. Recently, however, the comparative economics of alcohol fuels and gasoline have changed radically.

Methanol can also be used as the raw material to produce dimethyl ether ($(CH_3)_2O$), a completely clean-burning diesel fuel that could be used by trucks, locomotives, and ships. Many cars could also eventually use diesel. Diesel engines are substantially more efficient than traditional internal combustion engines, and equal to anything realistically possible from far more expensive—and as yet impractical—fuel cells.

As we shall show in chapter 8, by using these alcohol and alcohol-derived substances, we cannot only replace petrochemicals for fuels, but also create an alternative plastic and synthetic fabric industry that can produce virtually everything now made from petroleum products, but with less pollution.

FLEX-FUEL TECHNOLOGY

Two developments make a rapid transfer to high-alcohol fuels possible. One is the recent rise of gasoline prices, making methanol and ethanol economically attractive. The other is a technological innovation: the development of sensors capable of continuously measuring the suitability of the fuel-to-air ratio in cars using mixed alcohol/gasoline fuel, and using this information to regulate the engine.

With this breakthrough, some 6 million vehicles were produced between 1998 and 2006 that were capable of handling various alcohol/gasoline combinations. That is already five times the number of gasoline/electric hybrids on the road, and vastly increased use of such vehicles could happen overnight, for just a hundred dollars or so extra per vehicle (compared to many thousands more for hybrids).

The only sticking point is the nonavailability of high-alcohol fuel mixes at the pump. Filling stations don't want to dedicate space to a fuel mix used by only 3 percent of all cars. And consumers are not interested in buying vehicles for which the preferred fuel mix is extremely difficult to find.

This chicken-and-egg problem can be readily resolved by legislation. One major country has already done so. As we will discuss in chapter 9, in 2003, Brazilian lawmakers decided to engineer a national transition to FFVs, using tax incentives to help move things along. As a result, the Brazilian divisions of Fiat, Volkswagen, Ford, Renault, and GM all came out with ethanol FFV models in 2004, which accounted for 20 percent of the country's new vehicle sales that year. By 2006 about 70 percent of all new vehicles sold in Brazil were FFVs, producing significant fuel savings to consumers, a boost to local agriculture, and a massive benefit to the country's foreign trade balance.

ETHANOL OR METHANOL?

To date, all FFVs have been certified as either methanol/gasoline designs or ethanol/gasoline designs. Combined methanol/ethanol/gasoline FFVs have not yet been commercially produced. Their development poses only modest challenges, however. The question is, which alcohol would be the best one upon which to base our future alcohol-fuel economy?

Methanol is cheaper than ethanol. It can also be made from a broader variety of biomass material, as well as from coal and natural gas. Methanol is the safest motor fuel, because it is much less flammable than gasoline (a fact that has led to its adoption by car racing leagues).

On the other hand, ethanol is less chemically toxic than methanol and it carries more energy per gallon. Ethanol contains about 67 percent of the energy of gasoline per gallon, compared to 50 percent for methanol. Both thus achieve fewer miles per gallon than gasoline, but about as many miles per dollar at current prices, and probably many *more* miles per dollar at future prices.

Methanol is more corrosive than ethanol. This can be dealt with by using appropriate materials in the automobile fuel system. A fuel system made acceptable for methanol use will also be fine for ethanol or pure gasoline.

Both ethanol and methanol are water soluble and biodegradable in the environment. The consequences of a spill of either would be much less devastating than those of a spill of petroleum products. If the *Exxon Valdez* tanker had been carrying either of these fuels when it ran aground, instead of a vast oil slick, the environmental impact caused by the wreck would have been negligible.

Ethanol is actually edible, whereas methanol is toxic when drunk. This difference, though, should not be overdrawn, since in an FFV economy, both would be mixed with gasoline. The breakdown products of both ethanol and methanol are much less noxious than those from petroleum, and both emit far fewer air-polluting particulates when burned. Methanol and ethanol cars both produce less nitrogen oxide and ozone than gasoline vehicles. Since it is made exclusively from agricultural products, ethanol acts as a counter to global warming. Methanol can as well, but only if its source is biomass, urban waste, or stranded natural gas that would otherwise be flared (burned off). Methanol produced from coal or commercially viable natural gas has about the same impact on global warming as gasoline.

In short, either methanol or ethanol could be used very effectively, with roughly equal countervailing advantages. This has not stopped proponents of either fuel from vociferously arguing their unique advantage, and pushing for FFVs based exclusively on their favored product. To date, the more effective faction in this debate has been the ethanol group, backed as it is by the powerful farm lobby.

Given this political support, and no decisive technical argument in favor of methanol, the question might well be asked: Why not just go with the stronger side and implement an exclusively ethanol/gasoline FFV economy? The answer has to do with the total resource base. If we want FFVs not merely to benefit farmers, but also to make America energy independent, we need a larger production base than ethanol alone can deliver.

The United States uses 380 million gallons of gasoline per day. If we were to replace that entirely with domestically produced ethanol, we would have to harvest approximately *four times* as much agricultural output as we currently grow for food production. Now it is true that we don't need to replace all of our gasoline, at least not in the short term. Replacing half would make us substantially energy independent. Furthermore, future processes might eventually wring out higher ethanol yields per acre. Large supplies of surplus ethanol from Brazil and other tropical nations could also be imported. Nonetheless, relying on ethanol alone would require putting under fresh cultivation an amount of land greater than what we now use for food production. This would appear to be an unlikely prospect.

So if we are to use alcohol fuels to achieve energy independence, a broader resource base is needed. This can be provided by methanol, which can come from both a broader array of biomass materials and also from coal and natural gas. Methanol production from coal is particularly important, since coal is America's—and the world's—cheapest and most prevalent energy resource. The United States could power its entire economy on coal for centuries, and large reserves also exist in allied countries. Current coal prices stand at about three cents a kilogram, much cheaper than agricultural products, so methanol can be made from coal at low cost. By mixing it at various rates with ethanol over time, we can increase supplies, reduce prices, maximize environmental benefits, and vastly increase the flexibility of our alcohol economy. Furthermore, by making methanol from parts of farm products unsuitable for ethanol manufacture (for example, the stalks and leaves of corn plants), the economics of ethanol production can be significantly improved. Insisting that future vehicles have the capability to burn both alcohols is thus critical.

HELPING THE WORLD'S FARMERS

Even with methanol in the mix, the shifting of the world from a petroleum to an alcohol standard would remain a great boon to farmers. And third world farmers as much as American growers would enjoy the benefits—not only from a vastly increased market for their products, but also from the collapse of petroleum prices (which currently threaten crushing fertilizer and truck and tractor fuel prices). This adds a strong humanitarian case for the transition to flexible fuels.

It also adds to the strategic case. Currently, third world demagogues such as Venezuelan dictator Hugo Chávez are using the issue of advanced-sector import barriers to agricultural products as a red flag to seize power. Opening our markets would take this issue away.

By providing third world populations with an extensive source of income, the alcohol economy would also give them the wherewithal to buy manufactured products from developed nations. We would end up selling far more tractors and harvesters and hybrid seeds to Africans, for instance. That would improve the economic condition of all nations.

The extraordinarily positive results of such a policy for furthering global development, expanding trade, and raising living standards will be discussed in chapter 8.

THE WAY FORWARD

Energy conservation offers only a strained strategy for enduring economic oppression with very slightly ameliorated pain. Today's petroleum monopolists would still ultimately have us over a barrel. The ballyhooed hydrogen economy, meanwhile, is a hoax.

If we are to win the critical energy battle, there is only one way to do it. We must take ourselves—and the rest of the world—off the petroleum standard. Only by doing this can we destroy the economic power of our enemies at its very foundations. Only in this way can we transfer control of the future from those who *take* their wealth, premade, from the ground, to those who *make* their wealth through hard work, skill, and creativity (and thus must build free societies that maximize the human potential of every citizen).

Our nation's founders stipulated that the purpose of our government is to provide for our defense, promote our welfare, and secure the blessings of liberty to ourselves and our posterity. In our current economic and military dilemma, decisive action for energy independence is one of the most dramatic steps we could take to achieve those ends. Congress should immediately require that all future vehicles sold in the United States be flexible-fueled, thereby launching us into an alcohol-energy future that holds promise like few other options within our grasp.

In the next two chapters, we will discuss what will happen to us if they don't.

CHAPTER 2

TERRORISM

Your Gas Dollars at Work (Part 1)

> Imagine if the Ku Klux Klan or Aryan Nation obtained total control of Texas and had at its disposal all the oil revenues, and used this money to establish a network of well-endowed schools and colleges all over Christendom peddling their particular brand of Christianity. This was what the Saudis have done with Wahhabism. The oil money has enabled them to spread this fanatical, destructive form of Islam all over the Muslim world and among Muslims in the West. Without oil and the creation of the Saudi kingdom, Wahhabism would have remained a lunatic fringe.
>
> —Bernard Lewis[1]

Many Americans have been driven to consider the subject of alternative fuels and energy independence because they are angry about the high price of gasoline at the pump and prudently fearful that much worse may be soon to come.

While such concerns are entirely legitimate, there are larger questions involved with our dependency on foreign oil. For example: Who is getting our money? What are they doing with it? The several hundred billion dollars a year that Americans are shelling out for foreign oil represents an enormous amount of power—power to do good, or power to do evil. How is it being used?

By far the largest collector of international oil revenues is Saudi Arabia. The people of this country have been making a lot of headlines over the past few years. For example, fifteen of the nineteen September 11 mass murderers were Saudi subjects. Before that, it was Saudis who bombed the US Khobar Towers barracks in 1996, Saudis who bombed the US embassies in Kenya and Tanzania in 1998, and Saudis who organized the attempt to destroy the USS *Cole* in harbor in Yemen in 2000. Most of the "dead-ender" fanatics encountered in Afghanistan during the US post-9/11 counterattack were Saudis. Saudis have also comprised the largest group among the foreign terrorists slaughtering schoolchildren, attacking US troops, and creating general havoc for the purpose of inciting civil war in Iraq.[2]

But does this activity, carried out by a few thousand malignant individuals, have any relationship to the policies of the Saudi royals? After all, Timothy McVeigh's group came from Michigan, but no one blames the Oklahoma City bombing on Michigan's state government, clergy, or business elites. Perhaps a similar understanding attitude toward the Arab rulers might be appropriate as well.

Unfortunately, however, this is not the case. In point of fact, the Saudi ruling family has direct responsibility for promoting terrorism against the United States and many other nations. Furthermore, it is precisely the treasure obtained by looting the world of its fuel dollars that is allowing these people to expand their campaigns of terror, subversion, oppression, enslavement, corruption, and genocide all across the globe.

These are strong accusations. In this chapter I will prove them.

WHAT IS SAUDI ARABIA?

Saudi Arabia originated in the mid-eighteenth century as a partnership between Muhammad ibn Saud, the local bandit chieftain in control of the village of Diriyah in central Arabia, and Muhammad ibn 'Abd al-Wahhab, a religious firebrand enamored with the writings of the fourteenth-century Sunni Islamic scholar Ibn Taymiyyah.

Upset with the trends of his time, Ibn Taymiyyah argued that the Islamic world had become corrupted with such Christian customs as music, dancing, wine drinking, ornamentation of religious shrines, veneration of saints, socialization between the genders, and the excessive emancipation of women. Furthermore, he said that while the virtuous desert Muslims of old had correctly understood their duty to wage jihad as meaning an armed struggle against the infidels, their soft, modern, urbanized successors had falsely allegorized this obligation into mere religious introspection. Muslims, said Ibn Taymiyyah, had to return to the pure Islam of Muhammad and the early caliphs. These writings took root in various places across the Islamic world, giving birth to a number of "Salafist" movements.[3]

Ibn 'Abd al-Wahhab, however, added something extra to the basic Salafist doctrine. Since corrupted (i.e., non-Salafist) Muslims were not true Muslims, he said, armed jihad against them was as permissible and necessary as that against any other nonbelievers. Not only that, since non-Salafist Muslims engaged in the veneration of saints, they were actually not merely infidels, but polytheists (*mushrikum*). Therefore, according to the Koran, which states "kill the idolaters wheresoever ye find them" (sura 9:5), it was the duty of true Muslims to not merely subdue them, but to exterminate them.

This last distinction is important. Traditional Islam divided humanity into three groups: Muslims, infidels, and polytheists. Muslims were to wage jihad against the other two groups, but once conquered, the infidel "peoples of the book"—including Christians, Jews, and Zoroastrians—were to be allowed to live, albeit in the inferior social status of dhimmis, or "protected ones," who had to pay extra taxes and were forbidden to own arms, ride horses, bring legal actions against Muslims, build new temples, synagogues, or churches, or repair old ones. They also had to wear humiliating yellow identifying strips on their clothing to mark them out for free abuse, a custom that was later revived against the Jews in some parts of late medieval Europe and in Nazi Germany. If the dhimmis were caught proselytizing their faiths to Muslims, the penalty was decapitation. Nevertheless, provided they were adequately submissive, they were permitted to continue to exist and practice their faiths. Over time, it was felt, the pressure of living under such disadvantages would bring them around to accept Islam. In contrast, polytheists—such as pagans, Hindus, Buddhists, and Taoists—were to be killed on the spot.

In reclassifying insufficiently orthodox Muslims as polytheists, therefore, Ibn 'Abd al-Wahhab was making a giant departure from traditional Islamic practice. But furthermore, if such nominally Muslim people were beyond the pale, then clearly the Jews, Christians, and Zoroastrians had to be as well. Since this policy diverged explicitly from that ordained by the Prophet Muhammad himself, it was heresy. Wahhab justified himself, however, by saying that while the people of the book of the Prophet's time may have been monotheists, since then they had fallen from their faiths and become idolaters, sorcerers, and devil worshipers, deserving only of death.

Wahhab's vehement condemnations of the sins of various Arab chieftains made him an unwelcome guest, and he had to flee repeatedly to escape their revenge. Eventually, however, he was given refuge by the tribal leader Muhammad ibn Saud, and the two struck a deal: In exchange for Saud's protection and agreement to spread his creed by jihad, Wahhab would serve as his propaganda firebrand, rallying the Bedouins to his holy cause. To seal the covenant, Saud married Wahhab's daughter, and the Saudi royal family was launched.[4]

The alliance proved to be a great success. Wahhab's doctrine added fanatical bigotry to the Bedouins' own desire to murder and pillage, and under Saudi leadership, the tribes went on a rampage. Starting with a series of raids in the 1760s, they conquered central Arabia, and then branched out to subjugate Kuwait and Bahrain in the 1780s. In the face of this crazed menace, neighboring Muslims were dismayed. As one Omani chronicler commented at the time, "[Their doctrine] legalizes the murder of all Muslims who dissent from them, the appropriation of their property, the enslavement of their offspring, and the marriage of their wives without first being divorced from their husbands."[5]

In 1802 the Saudi-Wahhabis invaded Ottoman Iraq. They sacked the city of Karbala and destroyed its holy Shiite shrines, including the tomb of Hussein, the grandson of the Prophet Muhammad. In the ensuing slaughter, they murdered virtually the entire population of the town—men, women, children, and infants alike.[6]

But it was not only Shiites who felt the Wahhabi rage. In 1803 the Saudis took Mecca, massacred much of its orthodox Sunni population, and destroyed all the mosques, chapels, and shrines consecrated to the Prophet, his family, and many other Muslim saints and heroes. A year later they took Medina, desecrated its shrines, and plundered the tomb of the Prophet himself.[7]

The marauding did not stop there. Massacre followed massacre, and by 1810 the Wahhabi barbarians were raiding deep into Syria and Iraq, threatening both Damascus and Baghdad with rape and pillage.

Finally roused to the menace, the Ottoman sultan Mahmud II ordered his Egyptian governor, Muhammad Ali, to launch a counterattack into Arabia to retake Mecca and Medina. Apparently thinking prudence the better part of valor, Ali let his son Tusun lead the Egyptian invasion, which was crushed.

Ali, however, had more sons and more armies to send, and, persisting, succeeded in driving the Saudis out of Mecca and Medina in 1813. Finally, on *September 11*, 1818, Ali's eldest son, Ibrahim, took the Saudi capital of Diriyah, capturing the reigning Saudi king, Abdullah, in the process. Abdullah was sent to Istanbul, where the sultan had him beheaded in front of the main gate of the Hagia Sophia, along with one of his ministers and his imam. Since they were heretics, the Wahhabi leaders were denied Islamic burial. Instead, they were left on display for stray canine consumption for several days, after which the remains of their bodies were dumped into the sea. The sultan then ordered that prayers be offered throughout the empire to thank Allah for "the defeat of the adversaries of the Muslim religion."[8]

The Saudis, however, while down, were not out. Throughout the nineteenth century they continued guerilla warfare against the Ottomans, and consolidated their alliance with Wahhabism through further intermarriage between the two clans. Finally, when defeat in

World War I shattered the Ottoman Empire, the Saudis were once again unleashed.

Bursting out of their heartland stronghold, the Saudis lead a fanatical Wahhabi Bedouin horde called the Ikhwan on a sweep through Arabia and most of its neighbors during the 1920s, taking Mecca in 1924 and Medina in 1925. Again the orgy of murder of a century earlier was repeated, but this time on a much vaster scale. According to Palestinian journalist Said Aburish, "No fewer than 400,000 people were killed or wounded, for the Ikhwan did not take prisoners, but mostly killed the vanquished. Well over a million inhabitants of the territories conquered by Ibn Saud fled to other countries."[9]

While the Ikhwan had originated somewhat independently as a Wahhabi brotherhood among the Bedouins, Saudi king Ibn Saud used his funds and his stable of imams to take the movement over, and then radically expand it, turning its zealots into his favored shock troops. The Ikhwan, however, were determined to exterminate all people who they felt to be substandard, particularly the Shiites of the conquered eastern portion of the Arabian Peninsula, many of whom the king now wanted for his subjects. A conflict thus developed between Ibn Saud and his terrorists, and he was forced on several occasions to deploy his own tribal forces to discipline them. With the help of British air power, the king was able to prevail.

As the new regional imperial power replacing the Ottomans, however, the British had no interest in accepting the formation of a Saudi-Wahhabi empire dominating the Middle East. Much to Ibn Saud's displeasure, they forced him to accept a territory limited roughly to Saudi Arabia's current boundaries, assigning Transjordan and Iraq to two branches of the rival—non-Wahhabi—Hashemite clan instead. Other, smaller chunks of Ottoman territory were given to a variety of minor Arab potentates. The British kept Palestine, Egypt, and some other coastal principalities under their own control, while Syria and Lebanon were given to the French.

Taking what he could get, in 1932 Ibn Saud proclaimed his conquests a kingdom, naming it Saudi Arabia, after himself. For wealth, the new realm had its inhabitants, who were declared to be the personal property of its ruling family and who remain so to this day. The kingdom also had control of the holy cities of Mecca and Medina, which brought it Islamic prestige and some revenue from pilgrims, but little real power in the world at large. This situation began to change, however, in the late 1930s, when oil was discovered in the kingdom. Seeking to avoid excessive domination by the imperialistic British, the Saudis chose to make their oil development deals with the more strictly business-minded American oil companies instead. Thus began the Saudi-American alliance, which deepened during World War II, as the Roosevelt administration, convinced that domestic US crude was nearly tapped out, signed a treaty with Saudi Arabia to help ensure future fuel supplies.[10]

As an oligarchy of superrich oil despots lording over impoverished masses, the Saudi rulers became in the postwar years a natural target for both Communist and Arab national socialist demagogic propaganda. Moreover, such opponents were not all talk. In 1962, for example, the Soviet-supported Egyptian dictator Gamal Abdel Nasser invaded neighboring Yemen, thereby posing a direct threat to the kingdom. In the face of such menaces, the regime had little choice but to cling loyally to its American protector. Thus, despite joining with Venezuela as one of the two key initiators of the Organization of Petroleum Exporting Countries (OPEC) in 1960, the Saudis spent the 1960s acting as a moderating force on the nascent cartel's demands. As Wahhabi fanatics, the Saudis were ardent in spreading extreme anti-Jewish venom throughout the Muslim world. But aside from conspiring with other Arab states to keep the refugee Palestinians homeless for propaganda purposes, the Saudis shunned significant direct action in conjunction with the Soviet-backed alliance against Israel.

In 1972, however, the new Egyptian president, Anwar Sadat, expelled the Soviets from his country, thereby removing the primary external threat to the Saudi regime. Suddenly, American goodwill was no longer so essential, and the Saudis were quick to seize the opportunities the new situation afforded. In October 1973, using US support for Israel in the Yom Kippur War as a pretext, they declared an oil embargo against the United States and led OPEC in a unified action that sent petroleum prices up more than 700 percent.

This one move radically transformed the position of Saudi Arabia in the world. The kingdom's oil revenues, which had been a mere $2.7 billion in 1972, rose to over $22.6 billion in 1974. Nor was this the end of it. Using their newfound strength, the Saudis and most other OPEC members acted quickly to "renegotiate" their deals with the Seven Sisters international oil companies (Exxon, Mobil, Gulf, Chevron, Texaco, Shell, and British Petroleum), forcing the developers to hand over most of their previously agreed-upon share of the business. After some missteps, the Saudis learned how to use their position as the largest producer of the cheapest oil to discipline the other OPEC nations to stick to assigned production quotas, and then, controlling the cartel as a whole, alternate between artificially high oil prices over the long haul and intermittent sharp price drops to destroy would-be competitors. As a result, the incredible Saudi revenues not only persisted but continued to rise. In the three decades since the embargo, they have pulled in more than $2 *trillion*.

The Saudis have not used this money well. In 1974, nearly 91 percent of all Saudi income was from oil. Three decades and $2 trillion later, that percentage is unchanged.[11] There has been *no* real economic development in the kingdom. Instead, most of this vast treasure has been wasted on the most incredible collection of narcissistic expenditures the world has ever seen. Hundreds of palaces have been built; luxury cars, yachts, private jets, and thoroughbred horses bought and dissipated by the thousands; narcotics, fine liquors, jewelry, haute couture, child sex slaves, and concubines purchased and consumed by the ton. Through his twenty-two wives, and those of his forty-five sons, King Ibn Saud had more than six hundred grandchildren, and they, in addition to some three thousand other relatives from collateral lines, all needed to live royally, with allowances for each starting at $270,000 *per month*. Beyond these there were tens of thousands of additional "aristocrats," retainers, advisers, clerical allies, insiders, and other sycophants who were all dealt substantial cuts. Finally, there was the Saudi population at large, most of which was put on welfare, leaving the indignity of doing the kingdom's work to millions of indentured foreign coolies. (It is estimated that 70 percent of all jobs in Saudi Arabia, including 90 percent of all private-sector jobs, are done by foreigners. Only one Saudi in six works. The rest have their time free to spend idling in brothels or mosques.)

Still, even with all this waste, $2 trillion is a lot of money, and the Saudis had plenty left over to spend on their true passion: to wit, the spread of Wahhabi Islam throughout the world. This they undertook to do using a dense and complex assortment of nongovernmental organizations (NGOs); banks; charities; educational, religious, and cultural institutions; criminal gangs; and terrorist movements. Some of these originated before the 1973 oil embargo transformed Saudi finances, but all saw their power and reach radically expanded once the serious money started rolling in.

Some of the most important of these Saudi initiatives include the Muslim World League (MWL), the International Islamic Relief Organization (IIRO), the World Assembly of Muslim Youth (WAMY), and Al Haramain. We will discuss each of these in turn.

MUSLIM WORLD LEAGUE

Headquartered in the Saudi capital city of Riyadh, the Muslim World League (MWL) was founded in 1962 as the semiofficial organization for the global dissemination of Wahhabism. The head of the MWL is a Saudi government minister, its board is composed largely of the Saudi *ulama* (recognized religious leadership), and it is funded directly by the Saudi state. According to the Supreme Islamic Council for the United States, "it functions as the world headquarters for extremist Islamic networks."[12]

The MWL acts to spread and strengthen Wahhabism through both the establishment of new institutions and movements and the takeover of old ones, as well as through general preaching, publications, and propaganda. The initial personnel for its construction came from bringing together Saudi-Wahhabi clerics with the refugee leaders of the Egyptian Muslim Brotherhood. This latter Salafist organization had been founded in Cairo in 1928 by the religious scholar Hassan al-Banna, himself the student of the Wahhabi preacher, and paid Saudi royal family operative, Rashid Rida. The Muslim Brotherhood grew into a powerful movement, guilty of innumerable acts of terror and murder directed against Egypt's large Coptic Christian minority, Shiites, and secular Sunni Muslims. Unfortunately, the brothers' pious devotions did not stop there, but also included the prolific assassination of government ministers, first within the British puppet government of King Farouk, and then those of the secular national socialist dictator Gamel Abdul Nasser. When Nasser responded to this threat by expelling the brotherhood from Egypt, the Saudis welcomed them with open arms.[13]

Leading Muslim brothers, including Sheikh Omar Abdel-Rahman (later convicted for his role in the first World Trade Center bombing), future al Qaeda leader Ayman al-Zawahiri, Abdullah Azam, and Muhammad Qutb, were given not only sanctuary, but cushy teaching positions at Saudi universities. Thus brought comfortably into the fold, the brothers quite reasonably dropped their previous nitpicking objections to the irreligious hedonistic personal behavior (such as alcoholism) of the Saudi royals, and buckled down to the task of assisting the MWL team in its sacred mission of bringing all Muslims to embrace Wahhabism. As part of the MWL brain trust, the brothers poured out an endless stream of fanatical books and pamphlets for Saudi government publication and worldwide distribution, and helped shape the minds of hundreds of thousands of promising young scholars brought to Arabia from around the world to receive their instruction. This latter activity proved quite productive. From their positions teaching Islamic studies at King Abdul Aziz University in Jidda, for example, Azam and Qutb had the honor of educating Osama bin Laden.[14] In 1981 they were also able to organize the assassination of Egyptian president Anwar Sadat after Sadat committed the crime of making peace with Israel.

Thus, through incorporation into the MWL, the Muslim Brotherhood effectively became a branch of the Saudi-Wahhabi international. Another important Islamic fundamentalist movement that was taken over by the MWL in similar fashion was the Pakistan-based Jamaat Ulema Ilami (JUI). The point man for this effort was none other than Abdulla Azam, bin Laden's own mentor at King Abdul Aziz University. After finishing his tour there, Azam was sent by the MWL to Pakistan, where he ran their branch office and taught at the Islamic University in Islamabad. There he instructed thousands of local scholars, who in turn became the schoolteachers who would bring the Wahhabi doctrine of fanatical jihad to the impressionable young village boys of the Hindu Kush.[15]

Azam also identified the most radical existing JUI madrassas (local Islamic schools), and supplied them liberally with money, thereby creating an incentive for the rest to adopt extremism as well. Then, having gained a presence in the region, Azam proceeded to use copious Saudi funds to set up *thousands* of Wahhabi madrassas throughout the country.[16]

The hundreds of thousands of graduates of these madrassas soon developed into a fanatical terrorist movement that virtually destroyed civil society in much of Pakistan and murdered countless Hindus in the disputed province of Kashmir. On the Afghan border, they grouped together to create the horrific theocratic totalitarian force known as the Taliban.

In 1989 Azam was assassinated, probably by Soviet agents. His place as leader of the Taliban was briefly taken by his son-in-law, Abdullah Anas. A few years later, however, Anas was reassigned to head the Wahhabi Groupe Islamique Armé (GIA) killing spree in Algeria, and the Taliban franchise was passed on to Azam's student protégé, Osama bin Laden.

WORLD ASSEMBLY OF MUSLIM YOUTH

The World Assembly of Muslim Youth (WAMY) was established in 1972 as an initiative of the Saudi Ministry of Education for the purpose of spreading hard-line Wahhabism internationally.[17] WAMY receives Saudi government funds, and its operations are facilitated globally by Saudi embassies and consulates. It publishes and distributes Wahhabi texts around the world, with its top writers including Muslim Brotherhood terror theoreticians Sayyid Qutb and Muhammad Qutb and JUI jihad celebrant Sayyid Maududi. Hate tracts directed against Shiites and the alleged Jewish conspirators behind their unacceptable religion are also a staple of the WAMY literary effort. To round out their selections, WAMY's Saudi headquarters also publishes materials on bomb making, some of which were found in the homes of plotters of the 1993 World Trade Center attack.[18]

WAMY controls 450 Islamic youth and student organizations in thirty-four countries, including the United States. The primary purpose of these organizations is to spread Wahhabism among Muslim youth and to harass and suppress any anti-Wahhabi voices that might emerge from the Muslim student community abroad. In the United States, the leading WAMY affiliate is the Muslim Students' Association, but there are many others.

WAMY, however, does not limit its activities to campus politics. During the 1990s the president of WAMY in the United States was Abdullah bin Laden, brother of Osama, and under his leadership the organization allegedly maintained ties to al Qaeda.[19] In 1996 the FBI listed WAMY as an "organization suspected of terrorism."[20] Reportedly suspected of having contact with four of the September 11 hijackers,[21] the Washington, DC, office of WAMY was raided by federal agents in May 2004.

INTERNATIONAL ISLAMIC RELIEF ORGANIZATION

The International Islamic Relief Organization (IIRO) serves as a financial arm of the Muslim World League. It is fully funded by the government of Saudi Arabia.[22] In addition to financing many MWL proselytizing activities, it has also served as the conduit for transferring vast amounts of funds directly to a large portfolio of Wahhabi terrorist organizations created or inspired by the MWL.[23] These include the Abu Sayyaf organization in the Philippines, Hamas in Gaza and the West Bank, the Salafist Front Islamique du Salut (FIS) and Groupe Islamique Armé insurgents (guilty of an estimated 150,000 murders over the past decade) in Algeria, the Taliban in Afghanistan, and the al-Khattab terrorists operating against both the Russians and the pro-peace Sufi Chechens in the Caucasus.

AL HARAMAIN

While describing itself as a private charity, the Al Haramain Islamic Foundation is actually funded by the Saudi government and operates under the control of the Ministry of Religious Affairs. It deploys billions of dollars every year to promote religious extremism and terrorism around the world. Al Haramain operates in ninety countries; in forty of them, its offices are located in the Saudi embassy.[24]

Documents recovered by NATO investigators in early 2002 showed that the Bosnian office of Al Haramain was not only funding terrorist activities in the Balkans, but acting as a conduit for transferring Saudi funds to al Qaeda internationally.[25] Also implicated in the terrorist fund laundering was Al Haramain's Somalia branch. As a result, on March 11, 2002, the US Treasury froze the funds of these two organizations.

Al Haramain's terrorist portfolio, however, is much more extensive. Kenyan police report Al Haramain funding behind the attack on the US Embassy in Nairobi in August 1998.[26] Captured South Asian al Qaeda leader Omar al-Faruq has confessed that all the funding for his entire Indonesian organization came from Al Haramain.[27]

While the most prominent among the Saudi terror finance organizations, the IIRO and Al Haramain by no means exhaust the list. Also noteworthy are the DMI (Dar al-Mal al-Islamiyya, or House of Islamic Finance), the Organization of the Islamic Conference (OIC) and its associated Islamic Development Bank, the Islamic Benevolence Committee, the SAAR Foundation, the Holy Land Foundation, and many others.[28] In addition to these, direct private donations are made to terrorist groups by innumerable Saudi royal princes. Collectively, between the government-sponsored financial conduits and members of the ruling family operating in their private capacities, the Saudis are estimated to have spent several hundred billion dollars over the past thirty years promoting terrorist ideology and organizations, and directly funding operations. This is orders of magnitude more than was ever spent by the former Soviet Union and Red China, combined, over their entire history, to promote international communism.

Beyond this, there has been the open spending of the Saudi government itself, which has used its unlimited funds to buy almost every purchasable print and broadcast pan-Arab media outlet and to take over many of the most important universities in the Islamic world. This latter feat has been accomplished by bribing susceptible faculty members to join their cause, offering benefits such as paid sabbaticals in Saudi Arabia at twenty times their ordinary salary, and then using these paid stooges to intimidate, silence, or drive out the rest. Similar techniques have been used to take over mosques. Where mosques do not exist, the Saudis have built them, more than fifteen hundred worldwide, and staffed them with Wahhabi clerics to rally Muslims everywhere to the doctrines of total intolerance and murderous jihad.[29] They have also established some 210 Wahhabi Islamic Centers in cities around the world, including Malaga, Gibraltar, Madrid, Lisbon, Rome, Zagreb, Vienna, Geneva, Brussels, Mantes-la-Jolie, Lyons, Edinburgh, London, Brasilia, Rio de Janeiro, Tokyo, Toronto, Calgary, Ottawa, New York, New Brunswick (NJ), Washington, Toledo, Chicago, East

Lansing, Columbia (MO), Fresno, and Los Angeles. More than two hundred Wahhabi colleges have also been set up, with locations including Washington, Paris, London, Bonn, Moscow, and Bihać (Bosnia). The precise number of Wahabbi madrassas set up globally by the Saudis is unknown, but on the basis of partial data, it is estimated to exceed twenty thousand. These schools graduate hundreds of thousands of hate-indoctrinated fanatics every year.[30]

Between Afghanistan, Pakistan, Kashmir, India, Chechnya, Dagestan, Russia, Uzbekistan, Tajikistan, Algeria, Iraq, Israel, Egypt, Jordan, Lebanon, Turkey, Syria, Somalia, Kenya, the Sudan, Tanzania, Nigeria, Biafra, Bangladesh, Thailand, Indonesia, East Timor, Papua New Guinea, the Philippines, Bosnia, Serbia, Kosovo, Spain, England, and the United States, the death toll inflicted by this Saudi-fomented and financed international terrorist movement numbers in the *millions.*

The list of Saudi-Wahhabi terror campaigns is much too long to allow discussion of each of them here. However, it is worth taking a moment to talk about one of them, namely, the Afghan Taliban, because it shows what happens when the Wahhabis win.

The Taliban movement was a Saudi project. It is true that the Afghan resistance to the Soviet occupation was launched in 1980 on the initiative of Carter administration national security adviser Zbigniew Brezinski as a joint US-Saudi project, with equal monetary commitments from both parties. However, the Saudis went beyond funding arms shipments to the Afghan fighters opposing the Russians. They chose to start a war within the war by creating and launching their own fanatical Wahhabi shock force, and then, after the Soviets were run out of the country, they funded the Taliban march to victory over their Afghan tribal competitors. The government they then set up was a perfect Wahhabi utopia, featuring total degradation of women; the complete abolition of intellectual, religious, artistic, and political liberty; the end of meaningful education; the elimination of almost all activities, whether traditional or novel, that might contribute to joy in private life; and the minute regulation of every aspect of human existence by a theocratic totalitarian cult—all enforced in the most brutally murderous way imaginable.

This is the sacred cause for which the Saudis have launched their worldwide jihad. And your gasoline dollars are providing them the means to pay for it. But that's not the only thing they have been purchasing. They are also buying your government.

We discuss this very important issue in the following chapter.

CHAPTER 3

CORRUPTING WASHINGTON

Your Gas Dollars at Work (Part 2)

A secret team of American physicians follows the troops during their attacks . . . to ensure quick operations for extracting some organs and transferring them to private operation rooms before they are transported to America for sale. . . . These teams offer $40 for every usable kidney and $25 for an eye.

—*Al Watan*, official Saudi Arabian government newspaper
December 18, 2004

Pakistan and Saudi Arabia have become some of our most valuable allies in the war on terror.

—President George W. Bush, September 7, 2006

The Saudis have been looting our economy on a massive scale and are using the proceeds to fund a global war against civilization. One might think the United States would forcefully answer such a threat, but it has not. The Nixon administration did nothing about the 1973 oil embargo, and the American political class has preserved our oil vulnerability by failing to implement any effective energy policy to counter OPEC during the three decades since. The Clinton administration limited its response to the Islamist bombing of our embassies in Kenya and Tanzania to launching a few cruise missiles against some worthless facilities in the Sudan. President Bush replied to the Wahhabi slaughter of three thousand Americans in the September 11 attacks by striking a forceful blow against the Taliban, but he left the Saudi regime that created it—and all the 9/11 mass murderers—untouched. Why has our response been so weak?

There is strong evidence that the American reaction has been hamstrung by a sustained Saudi campaign to buy influence in our political system.

In his very interesting book *Sleeping with the Devil: How Washington Sold Our Souls for Saudi Crude*, former CIA counterterrorism case officer Robert Baer reports the following incident concerning the activities of Saudi royal family bagman Adnan Khashoggi, the son of Ibn Saud's personal physician:

> In late 1968, days after Richard Nixon won the White House, Khashoggi was one of the first to fly out to congratulate the president-elect. He didn't forget to pass on the regards of Interior Minister Fahd—the prince who sent him to San Clemente and the current brain-dead king. When Khashoggi got up to leave, he "forgot" his briefcase, which happened to be stuffed with $1 million in hundreds. No one said a word. Khashoggi went back to his hotel to wait for a telephone call. The phone never rang. It never would. A couple of days later, and Khashoggi knew the trick had worked: Washington was for sale. Like original sin, that changed everything.[1]

If this account is true, then the Saudis certainly received a great return on their investment five years later, when the Nixon administration rebuffed calls by the hawkish Senator Henry "Scoop" Jackson (D-WA) to seize the kingdom's oil fields to break the Arab embargo.[2]

However, while many of such cloak-and-dagger, behind-the-scenes Saudi illegal bribery stories may well be factual, they are not necessary to explain how the Wahhabi plutocrats have warped America's political processes. With so much largesse to deploy, legal means can be entirely adequate.

Let's start with the most open and aboveboard Saudi influence peddling. The Foreign Agents Registration Act of 1938, still on the books, requires agents of foreign powers operating in the United States to declare themselves, and some actually do. The list of these self-declared agents is then presented to Congress by the Justice Department every six months. For the first six months of 2005, the officially listed Saudi agents were: Barnett Group, LLC; Dutton & Dutton, PC; Gallagher Group, LLC; Hill and Knowlton, Inc.; Loeffler Tuggey Pauerstein Rosenthal LLP; MPD Consultants, LLC; Patton Boggs, LLP; Qorvis Communications, LLC; Sandler-Innocenzi, Inc.; Saudi Petroleum International, Inc.; and Saudi Refining, Inc.[3]

This is a startling list. Patton Boggs is one of the most powerful lobbying organizations in Washington, DC. It is a political nine-hundred-pound gorilla with four hundred attorneys and legions of legal auxiliaries on its staff. It owns Qorvis Communications. According to the Justice Department report, these two together accepted payments of $5,845,106.36 from the Saudi government during the first six months of 2005 alone.

Sandler-Innocenzi received almost $1.6 million for the period, and, according to the report, "placed radio buys and print advertising on behalf the foreign client through Qorvis Communications," so the Patton Boggs take may have been even larger.

Loeffler Tuggey Pauerstein Rosenthal is a lobbying organization led by former House majority whip Tom Loeffler (R-TX). Loeffler was the national finance cochairman for the George Bush 2000 election campaign.[4] His company received $734,062 from the Saudis for the period.

But these are just the people who admit they are on the Saudi payroll. There are innumerable other influentials who accept well-paid consultancies from the Saudis and who chose not to make the connection public. One of these appears to be former secretary of state Henry Kissinger, who had to resign from his position as head of the September 11 investigative commission when he was asked to disclose his client list. Kissinger wasn't alone. Former Senate majority leader

George Mitchell (D-ME) had to quit as commission vice chair for the same reason.[5]

Over the past two decades, the Saudis have spent about $100 billion in arms purchases, much of which has gone for highly sophisticated equipment they are unable to use. Nevertheless, by judiciously placing these purchases, they have been able to mobilize armies of lobbyists from major defense contractors to spout the kingdom's line. Since such contracts translate into jobs, and thus votes, they have served as a powerful means of influencing politicians as well.[6]

Then there are the business partnerships, financed by Saudi funds, which invite influential Americans (and Europeans) to spend a little time on their boards in exchange for fantastic compensation. The most famous of these is the Carlyle Group, a global investment firm with $56 billion under management, whose equity stems largely from Saudi investors, including Saudi Arabia's longtime (just-retired) US ambassador Prince Bandar bin Sultan; his father, Saudi Arabian defense minister Prince Sultan bin Abdul Aziz (himself the son of Ibn Saud and brother to the king); and the bin Laden family. In addition to such luminaries as Shafiq bin Laden, the brother of Osama bin Laden, and Saudi prince Al-Walid bin Talal (ranked as the world's eighth richest person, worth $20 billion, according to *Forbes*, in 2006), Carlyle's leadership includes former secretary of defense Frank Carlucci, former secretary of state James Baker, former SEC head Arthur Levitt, former FCC chairman William Kenard, former World Bank treasurer Afsaneh Beschloss, former Clinton administration White House chief of staff Thomas "Mack" McLarty, former Office of Management and Budget (OMB) director Richard Darman, former British prime minister John Major, and former president of the United States George H. W. Bush.[7] In 1990, future president George W. Bush was given a position on the board of Carlyle subsidiary Caterair.[8]

According to the March 5, 2001, *New York Times*, James Baker's gains from his participation in Carlyle were estimated at $180 million. It pays to be a friend of the Saudis. In 2003 Baker's law firm, Baker Botts, LLP, was hired by the Saudi Arabian government to defend it against the legal action initiated by the relatives of those killed by the Wahabbis' September 11 attack.[9]

If you are looking for some spare change, being a member of the Saudi favorite list can pay out in other ways. For example, during the 1990s the Gulfstream Aerospace Corporation wanted to make some luxury jet sales to Saudi Arabia and Kuwait. So they invited Donald Rumsfeld, Colin Powell, Henry Kissinger, and former secretary of state George Schultz onto their board, giving them low-cost stock options and the assignment of opening the right doors. The Saudis obliged their old friends, and, delighted with the results, Gulfstream allowed them all to cash out their options. According to the December 21, 1998, *New Republic*, Kissinger made $876,000 for five months' work on this deal, Schultz got $1.08 million, Rumsfeld $1.09 million, and Powell $1.49 million.[10]

As Saudi ambassador Prince Bandar explained, "If the reputation then builds that the Saudis take care of friends when they leave office, you'd be surprised how much better friends you have who are just coming into office."[11] Bandar was a particularly effective ambassador, because his father, Prince Sultan, the Saudi defense minister, is the one who dishes out the multibillion-dollar Saudi military purchase contracts.[12] Be good to Bandar and he can be very good to you.

Beyond such cozy personal deals, Saudi influence spreads much further through conventional business partnerships. For example, Chevron Texaco is a partner with Saudi Aramco in both Star Enterprise and Motiva Enterprises. Chevron Texaco is also engaged in joint enterprises with Nimur Petroleum, which is owned by the US-court-indicted Saudi bin Mahfouz family, noteworthy as major operators in the Bank of Credit and Commerce International (BCCI) criminal finance syndicate.[13] Chevron's board members include former HUD secretary and US trade representative Carla Hills, former Louisiana senator J. Bennett Johnson, former Georgia senator Sam Nunn, and, prior to her current government stint, Secretary of State Condoleezza Rice. Amerada Hess is involved in joint business ventures with several members of the Saudi royal family. Amerada Hess board members include former New

Jersey governor (and September 11 investigation commission chairman) Tom Kean, former Treasury secretary Nicholas Brady, and George H. W. Bush special assistant Edith Holiday.[14] Other Washington power players with business interests linked to the Saudis include former CIA director John Deutch, former national security adviser Brent Scowcroft, former secretary of state Lawrence Eagleburger, and, until his recent death, Lloyd Bentsen, former Texas senator and Democratic candidate for vice president in 1988.[15]

The Saudis also like to hire law firms with politically connected partners. In addition to Patton Boggs and Baker Botts, the kingdom has also sought representation from Akin Gump, a firm whose partners include Democratic Party heavyweight fixers Vernon Jordan and Robert Strauss, and former Speaker of the House Tom Foley (D-WA).[16] Akin Gump, however, has not limited its representation to the Saudi government. The firm has also served as legal counsel to the Hamas-linked Holy Land Foundation, and to Khalid bin Mahfouz, Mohammed Hussein Al-Amoudi, and Salah Idris—all of whom have been scrutinized or indicted by US authorities for possible involvement in financing al Qaeda.[17]

As Baer puts it, "At the corporate level, almost every Washington figure worth mentioning has served on the board of at least one company that did a deal with Saudi Arabia."[18] Surely not all of these people are defined by such involvement. Yet business is business, and under these circumstances one would be naive not to suspect the possibility that the financial self-interest of at least some of these enterprising individuals could be playing a highly inappropriate role in the US government decision-making process.

But not everyone can be corrupted by material inducements. The Saudis have therefore made substantial donations to virtually every major Washington think tank, the better to influence the American political class by controlling its ideas at their source. In addition, they have financed important policy institutes that are virtually their own, including the Meridian International Center and the Middle East Institute in the nation's capital, as well as additional centers at Harvard Law

School, Howard University, Syracuse University, Shaw University, American University, Duke University, Johns Hopkins University, the University of Arkansas, the University of California at Santa Barbara, and Georgetown University, America's school for career diplomats.[19]

The Saudis have also sought to influence US public opinion and make friends in the press through the hefty placement of advertisements. However, if you want a friendly press, there is no better way to get it than to own it. In 2005 Saudi prince Al-Walid bin Talal began the acquisition process by buying a 5 percent share in AOL-Time Warner, which owns CNN.[20] He also bought 5.6 percent of the voting shares in Rupert Murdoch's News Corporation organization, which owns Fox.[21] Al-Walid's other investments include a recent $27 million donation to a Saudi telethon conducted to raise money for suicide bombers.[22] In November 2005 he had a chance to begin to exercise some of his newfound media influence by calling Murdoch to "ask" that Fox's coverage of the Muslim riots in France be "changed." It promptly was.[23] On the evening on July 13, 2006, CNN's John Snow, substituting for Anderson Cooper, began his *360* show with the words, "People are paying a high price at the pump for Israel's war on terror." This is the Saudi line. Americans can expect to hear much more of it as the prince's media acquisitions expand.

Last, but not least, the Saudis have sought, not unsuccessfully, to warp US foreign policy through direct social fraternization with America's elites. With unlimited wealth to burn, the Saudis throw great parties at their palaces and estates around the world, complete with unlimited narcotics, top-of-the-line prostitutes, and the opportunity for attendees to mix with the rich and powerful, movie stars, supermodels, rock sensations, sports figures, and all the rest. Apparently, for some people, this sort of thing is quite a draw.[24]

Others prefer less dissolute pleasures, such as safaris to exotic locations with the Saudi royals. Colin Powell used to stay in touch with Prince Bandar by serving as his frequent racquetball partner.[25] In 1990 Bandar's US residence wife, Princess Haifa (daughter of Prince Sultan's brother King Faisal, and thus Bandar's first cousin), was kind enough to invite then-president George H. W. Bush's daughter, Dorothy, to celebrate Thanksgiving with her on the Bandars' Virginia estate. The president returned the favor by having Bandar vacation with him in Kennebunkport, after which he gave him the affectionate nickname "Bandar Bush." Delighted, Bandar invited the president to hunt pheasant with him on his estate in England, and for good measure, donated a million dollars to build the Bush Presidential Library in College Station, Texas.[26] Bandar has been quite generous to the Carters and Clintons as well.[27]

As president, George W. Bush has hosted Bandar at his ranch in Crawford, Texas. No other ambassador—not from America's closest ally, Britain; our top trade partner, Canada; nor our leading military or economic peers, Russia or Japan—is ever accorded such intimate relations with the US head of state.[28] One really has to wonder about such hospitality being offered to the official representative of a family that has financed tens of thousands of schools worldwide to instruct boys to kill Americans.

In late November 2002, *Newsweek* magazine exposed the fact that Princess Haifa had given $130,000 of her own money to an intermediary who passed it on to a number of Saudis operating in the United States, including two of the September 11 hijackers.[29] Such a sum amounts to about 25 percent of the total that the FBI estimates was required to pull off the 9/11 attack. Undeterred, Barbara Bush (the mother of the sitting president) and Alma Powell (the wife of the then-sitting secretary of state) both called Princess Haifa to offer her their full sympathy and support.[30]

Can America be defended by elites whose loyalties are so confused?

The Saudis are looting our economy to the tune of trillions of dollars. They are using a portion of these proceeds to finance a worldwide jihad against civilization, while deploying another part to corrupt our political system so that we don't strike back. If this situation is not corrected, it will soon grow much worse. With the industrialization of China and India, demand for oil is growing rapidly—a development that will set the stage for skyrocketing prices and consequent exponential further growth of Saudi power.

The Saudis are not the only ones using oil revenues to finance terrorism. Iran is another very active player. One of its favorite terrorist groups is Hezbollah. In addition to frequently attacking Israel with missiles, Hezbollah has been waging a much more deadly war against the West through massive heroin shipments to Europe.[31] Hezbollah has also recently established bases in South America and Southeast Asia. In Latin America it has linked up with Columbian narcoterrorists, supplying them with arms in exchange for cocaine.[32] Hezbollah also has established Islamist terrorist training camps in Paraguay, whose graduates may soon pose a significant national security concern for the United States. Iran has also bought its way into a minority share of control of the Wahhabi Muslim Brotherhood–initiated Hamas terror movement.

While their open hostility to the United States makes it difficult for the Iranians to engage in the Saudi game of corrupting Washington, they have been able to use their money to buy influence in the Russian and Chinese capitals instead.[33] This has occurred despite the fact that Iran has been sponsoring murderous anti-Russian terrorist groups in many of the former Soviet republics of central Asia[34] and also causing a catastrophe within Russia itself though massive running of heroin and other hard drugs into the country.[35]

The Iranians are Shiites, and thus viewed by the Wahhabi Saudis as candidates for extermination. Accordingly, the two Islamofascist powers are enemies, and each has sought and bought the great power of military and diplomatic allies for support against the other. But that has not stopped both from operating globally to foment hatred and terror against West and East alike. Indeed, rather than oppose each other's terror operations abroad, their relationship appears to be one of ardent competition to show who can do the most to destroy the common infidel foe.

In addition, however, the Iranian fanatics are using their petroleum revenues to build atomic bombs. The United States would like very much to stop this program, but the petrodollar-purchased Iranian fifth column in Moscow has induced Russia to block the imposition of any international sanctions that might halt it peacefully. Thus we may soon face the horrifying choice between nuclear-armed terrorists or superpower confrontation.

The issue of our energy dependence is *not* a mere matter of being forced to pay $3 per gallon for gasoline. Both the American republic and the future course of human civilization are at stake. We need to break the oil cartel before it is too late.

In the next chapter, we discuss how this can be done.

FOCUS SECTION: SAUDI INTERVENTIONS INTO US POLICY

With so many Saudi allies settled among the American political class, it would not be surprising if instances were to be discovered where our government's decision-making processes were warped to serve Saudi interests. In this section we will discuss a number of important incidents in recent US history in which improper Saudi intervention may have played an unfortunate role.

A. The Case of the Shiite Massacre

There is no question the Saudis were forcefully in favor of US intervention against Saddam Hussein after his invasion of Kuwait in 1990. Not only did the house of Saud ask for American action, they paid for it, to the tune of some $30 billion.[36] That said, few would fault the first Bush administration for supporting the Saudis in this matter, since repelling the Iraqi dictator's attempt to seize control of the world's largest oil supplies was clearly in America's interests as well. Yet after Saddam's forces were mostly defeated, the George H. W. Bush team made a sudden decision to stand down all US and coalition forces, thereby allowing the dictator to massacre the Shiites and Kurds who had risen up to fight by our side at the president's call. The immorality of this betrayal exceeds words. Indeed, to find its like one must depart from American history entirely, and compare it to the infamous decision of Communist dictator Joseph Stalin, who stopped the Red Army in its tracks at the gates of Warsaw in 1944 in order to allow the Nazis time to wipe out the rising of the Polish underground.

As paymasters for the first Gulf War, and patrons of such members of the Bush team as Joint Chiefs of Staff Chairman Colin Powell, Secretary of State James Baker, National Security Advisor Brent Scowcroft, and the president himself, there is little doubt that the Saudis had the power to call the shots on this matter. Why would they do so? Various analysts have presented complex geostrategic reasons for the Saudi decision,[37] and there may well be some truth in their arguments. However, asking why the Wahhabis supported the massacre of Iraqi Shiites in 1991 is like asking why the Nazis supported the slaughter of the Jewish population of Kiev in 1941. It is an article of faith for the Wahhabis that the Shiites should be exterminated, and no additional reason to explain any particular atrocity is required.

As a result of the Bush I betrayal, some three hundred thousand Shiites and tens of thousands of Kurds who had placed their trust in the United States were massacred, and a hideous stain was put on the American flag. A decade of sanctions costing further hundreds of thousands of Iraqi lives and tens of billions of US dollars was made necessary, and the task of deposing Saddam Hussein was left to the president's son, to be accomplished at a cost of several hundred billion dollars, more than twenty-six thousand American troops wounded, and more than thirty-eight hundred soldiers and marines dead.

B. The Case of Yasser Arafat

Contrary his own and Israeli propaganda, the late Yasser Arafat was not really, or at least primarily, a terrorist. Rather, he was a full-time criminal who engaged in part-time terrorism for public relations purposes. Some may find this a surprising assertion, but it is documented at length, for example, by former Drug Enforcement Agency investigator

Rachel Ehrenfeld in her very informative book *Funding Evil*.[38] Indeed, according to both 1990 CIA and 1994 British National Criminal Intelligence Services reports,[39] Arafat's Palestine Liberation Organization (PLO) is one of the largest drug-smuggling rings in the world, doing over a billion dollars per year of heroin traffic into Europe. The PLO also engages in bank robberies, arms trafficking, human trafficking, money laundering, protection rackets, and many other scams. One of its favorites is car stealing and resale, an activity that it conducts on an industrial scale, with the primary victims being Palestinians. Arafat's wife, Suha, owns the factory that produces the required counterfeit license plates. As a result of such entrepreneurial activities, the PLO has socked away an estimated $10 to 14 billion in Switzerland.[40]

While not a Wahhabi himself (although he was a fellow traveler of the Muslim Brotherhood in his youth), Arafat enjoyed consistent support from the Saudis, as well as from the secular fascist Arab dictatorships. Arafat earned this support by murdering any Palestinian alternative leadership inclined toward peace with Israel, launching intifadas, and using pinprick terrorism to provoke Israeli counterstrikes on the Palestinian people. Such actions helped the Wahhabi (and secular fascist regime) agendas by keeping the conflict going, thereby ensuring that the Palestinians would be kept perpetually immiserated for agitational purposes.

The Israelis would have killed Arafat long before his time if not for restraint imposed upon them by the US government. This restraint, in turn, was dictated to successive US administrations by the Saudis. An example of how this worked is reported by Baer: "In April 2001 Yasser Arafat called Saudi Crown Prince Abdullah to complain after Israeli soldiers fired on a convoy ferrying officials of the Palestinian Authority. . . . Abdullah in turn called Bandar, who called Dick Cheney, who called Colin Powell. . . . Within an hour of Arafat's call to Prince Abdullah, Powell was reading the riot act to Ariel Sharon in Tel Aviv. Tinkers to Evers to Chance was never so efficient."[41]

It is this sort of protection that allowed the Arafat criminal syndicate to operate for years.

C. The Case of the Neutralized Opponents

Those among the US political class who cross the Saudis have a way of finding themselves neutralized. Some of these neutralizations may have cost the nation dearly. We briefly consider the cases of four key individuals who might have caused serious problems for the Saudis in the Justice Department, the CIA, the FBI, and the Defense Department, had they not been neutralized in a timely fashion.

Kimba Wood is a prominent New York US District judge. In 1992, federal investigators presented her with evidence showing involvement of Saudi royal family money handler Khalid bin Mahfouz and his Bank of Credit and Commerce International (BCCI) in channeling large sums of money to terrorists and other criminal operations. Wood responded very forcefully, freezing all Mahfouz-owned or controlled funds, "assets, or other property . . . within the jurisdiction of the United States."[42] In early 1993 the Clinton administration, looking for a well-qualified female Democrat to fill the position of attorney general, offered Wood the job. Shortly after she accepted, however, a nannygate character assassination scandal operation was run against her, conducted from inside the Clinton camp, and her nomination was then axed under the insistence of White House Assistant General Counsel Bruce Lindsey.[43] Subsequently, Lindsey was given a position as an associate with Akin Gump, the Washington, DC, powerhouse law firm that represents Mahfouz.[44]

Wood was forced to withdraw. Instead, the position of attorney general was given to Janet Reno, under whose guidance the Justice Department did very little to stop al Qaeda's US domestic preparations for September 11.[45]

In early 1993 James Woolsey, a former undersecretary of the navy in the Carter administration, was given the position of CIA director by Bill Clinton. Woolsey was (and is) deeply anti-Saudi, but was unable to do anything effective about the Wahhabi threat because, despite being CIA director, he was denied all access to the president.[46] This is incredible, since the CIA is supposed to be the president's eyes and ears. Woolsey was told that the reason he was denied access was that Clinton "was not interested" in foreign policy, but this was manifestly untrue. Despite his preoccupation with healthcare and other matters of domestic politics, Clinton was passionately concerned about developments in the former Soviet Union, an area where the CIA had unmatched expertise, as well as in trying to somehow arrange a Mideast peace. During Woolsey's term as director, Clinton also engaged in foreign military operations in Bosnia and Somalia, so he had plenty of need of foreign intelligence. Yet Woolsey was given no access, and, disgusted with being kept on the shelf, resigned after two years in office. His position was then given to John Deutch, who stripped the agency of most of its covert human intelligence assets.[47]

The mystery of Woolsey's neutralization becomes less obscure if one considers who it was that controlled access to the president. During Woolsey's tenure at the CIA, the person in question was White House Chief of Staff Thomas "Mack" McLarty. McLarty is a partner of Kissinger Associates and a member of the Carlyle Group.[48]

John O'Neill was the head of counterterrorism for the FBI. Dedicated to his job, O'Neill vigorously pursued the evidence behind the bombing of the USS *Cole*, and tracked it back to a Saudi businessman with links to al Qaeda operating in Yemen. O'Neill wanted to go to Yemen to pursue the matter, but was blocked from getting a visa by unknown bureaucrats in the Clinton State Department. He was then further sabotaged by FBI management, who chose to leak his personnel file to the press. Disgusted, O'Neill resigned his government job and took a position as security chief of the World Trade Center. He was killed on September 11, 2001.[49]

Laurent Murawiec was a RAND Corporation analyst who presented an extensive dossier of evidence showing Saudi involvement in financing international terrorism to the Defense Policy Board (DPB) on July 10, 2002. This was a very important briefing, because the DPB reports directly to the secretary of defense. Its conclusions were unambiguous: "The Saudis are active at every level of the terror chain, from planners to financiers, from cadre to foot soldier, from ideologist to cheerleader." In response, DPB member and RAND trustee Frank Carlucci had Murawiec fired from his post at RAND. Carlucci was chairman of the Carlyle Group.[50]

D. The Case of the Missing bin Ladens

Shortly after the September 11 attacks, a special plane was chartered by arrangement with the White House that whisked more than a hundred prominent Saudis, including many members of the bin Laden family, out of the United States before they could be questioned by the FBI. This notorious incident has been the subject of much comment, including Michael Moore's movie *Fahrenheit 9/11* and Craig Unger's book *House of Bush, House of Saud*.[51] Although these sources are highly suspect because of their extremely partisan character, the incident nevertheless really seems to have happened,[52] and if so, is indicative of a very serious problem at the top.

E. The Case of the Missing Weapons of Mass Destruction (WMDs)

In March 2003 the second Bush administration launched the US invasion of Iraq. The administration justified the invasion by arguing that the Iraqi weapons of mass destruction (WMD) program was a threat to the world. In addition, Iraqi dictator Saddam Hussein was a genocidal fascist dictator who deserved to be overthrown for the sake of human rights.

The first of these reasons turned out to be false: There were no WMDs. (This is not too surprising—if there had been, an invasion would have been unthinkable.) The second rationale was true enough, and, in my view, defines the US war effort in Iraq as a moral enterprise. As a practical matter, however, one must ask why his crimes against freedom identified Saddam Hussein in particular as the top priority target for invasion, when there were plenty of worse offenders around—including Saudi Arabians.

Leftist critics of the war have declaimed it as a struggle for oil. But this theory is indefensible, since if the Bush crew really wanted to grab oil, Saudi Arabia would have been a far more lucrative place to go. Moreover, such a seizure would have been arguably justifiable, since, unlike Saddam Hussein, the Saudis had, in fact, been using their oil wealth to promote devastating attacks upon America.

So why did the Bush administration attack Iraq? Was it just the son trying to set right the error of his father? Maybe. But let us consider: who benefited most from the action? To what nation was Saddam Hussein the greatest threat? Answer: Saudi Arabia. In fact, it was the threat posed by Saddam Hussein that forced the Saudis to accept US troops on their soil. Such forces posed a serious concern for the Saudis, because they were a potential springboard for an American seizure of their oil fields. But with Saddam gone, the Yanks could be sent packing, too.

By taking down Iraq, we dramatically weakened our leverage with the Saudis. Furthermore, in the postinvasion chaos, Wahhabi terrorists have had a field day killing Shiites (as well as US troops), and have also done wonders for the kingdom's bottom line by keeping Iraqi oil off the world market.

You have to hand it to Prince Bandar.

F. The Case of the Sudan Genocide

For the past decade, the Saudi-backed Wahhabi Arab government of the Sudan has been waging jihads of mass extermination, first against the Christian and animist African population of the nation's south, and more recently against the black non-Wahhabi Muslims of the western region known as Darfur.[53]

Bush administration spokesmen, including Secretary of State Condoleezza Rice and President Bush himself, have correctly branded this activity as genocide, but have done nothing to stop it. Instead they have chosen to go whining and wailing to the UN, begging it to act. And this is despite the fact that they know in advance that China, itself on the hook for Sudanese oil, will block any resolution for sanctions, let alone effective action.[54]

I have the greatest respect for the Doctors Without Borders and similar groups trying to do something to help the people of Darfur. But the fact of the matter is that armed genocide cannot be stopped by medical care. Military force is required.

The Darfur genocide is being perpetrated by government-backed Arab Janjaweed militia traveling on horseback or in pickup trucks, supported occasionally by the few obsolete aircraft of the Sudanese air force. These airplanes could all be destroyed on the ground by a single strike of carrier-based fighter aircraft or cruise missiles launched from the Gulf of Aden, after which the Janjaweed could be substantially smashed by a squadron of Diego Garcia–based B-52s, armed with GPS-guided JDAM bombs, patrolling the country for a week or two at 45,000 feet. Once that is done, the Darfurians will be able to defend themselves. Not a single American soldier needs to set foot in the country.

George Bush could stop the Darfur genocide with a phone call to the Pentagon switchboard. Morality clearly demands that he do so, and under the Genocide Treaty of 1948 (as well as because the Sudanese government has committed acts of war against the United States by supporting al Qaeda's bombing of our African embassies), the legal authority for the use of force is there as well. Yet he does not act. Why not?

G. The Case of the Missing Energy Policy

The most egregious piece of pro-Saudi fecklessness of the American political class, however, has been its failure to enact a competent energy policy to stop the Arab looting of our nation. Some of the other disasters described above might be excused as mistakes due to stupidity or various personal factors. But the energy issue has been with us now for a third of a century. It is time we dealt with it.

CHAPTER 4
HOW TO DESTROY OPEC

Oil prices are expected to decline briefly (through 2006), then rise by about 0.7 percent per year to $27 per barrel in 2025.

—*International Energy Outlook, 2004,*
US Department of Energy, April 2004

The ceiling for the oil price should be infinite. . . . The American empire will be destroyed . . . Inshallah.

—Venezuelan president Hugo Chávez,
opening address to the OPEC conference, June 1, 2006

The power of the Saudis to foment global terrorism rests upon one thing: their copious supply of money. Yes, it is true that they also lead the worldwide Wahhabi movement and possess strings of political insiders to divert any counterattack that might come from the West. But until the Saudis started racking up tens of billions of dollars in inflated oil revenues, their cult was regarded by Muslims everywhere else as little more than primitive insanity. It is their unlimited funds that have allowed the Saudis to buy up the faculties of the Islamic world's leading intellectual centers, to build or take over thousands of mosques, and to establish tens of thousands of Wahhabi madrassas, pay their instructors, and provide the free daily meals necessary to entice legions of poor village boys to attend. Without rivers of treasure to feed its roots, this horrific movement could neither grow nor thrive. As for the Saudis' well-placed guardians within the American and European political elites, their loyalty is strictly contingent on continued largesse. The distinguished gentlemen of the Carlyle Group may be a lot of things, but one thing they are not is committed Wahhabis. They are in it for the money. Cut the gravy and they are off the train.

To paraphrase the famous aphorism of an otherwise forgotten war minister of France's Louis XII, there are three things necessary to wage jihad: money, money, and yet more money.[1]

The overwhelming source of Saudi income is oil. Since 1973 their petroleum revenue stream, and the power that goes along with it, has been fabulously multiplied by a cartel arrangement known as the Organization of Petroleum Exporting Countries.

To defeat the Saudis, we need to destroy OPEC. Let's see how this can be done.

HOW OPEC WORKS

In order to wreck OPEC, it is first necessary to understand how it works.

OPEC is an open conspiracy, in which representatives of the rulers of eleven kleptocracies (Algeria, Indonesia, Iran, Iraq, Kuwait, Libya, Nigeria, Qatar, Saudi Arabia, United Arab Emirates, and Venezuela) get together at periodic meetings and decide what the world price for oil should be, and then assign production quotas to each so as to force the price to that level. This is very different from the way business is conducted in a free market, and it produces very different results.

In November 1999 Saudi oil minister Ali al-Naimi told the Houston Forum that his country's "all inclusive" cost of producing petroleum was $1.50 per barrel, and its cost for discovering new reserves was

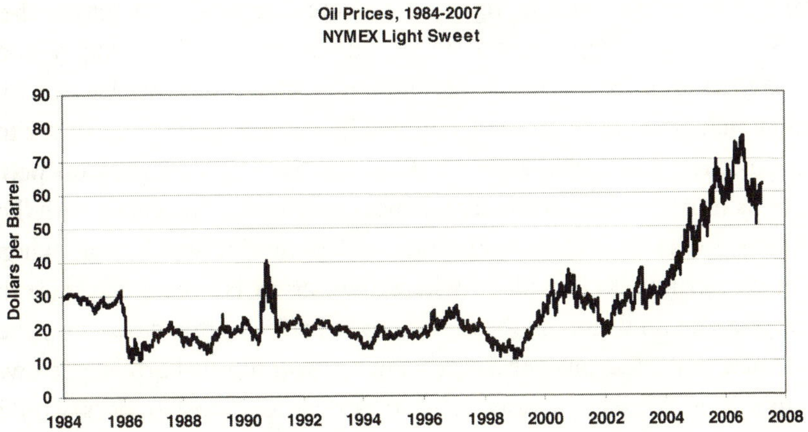

Oil Prices, 1984-2007
NYMEX Light Sweet

Figure 4.1. Actual (non-inflation-adjusted) price of oil, 1984–2007.
***Source*: Energy Information Administration, http://www.eia.doe.gov.**

about $0.10 per barrel.[2] Worldwide, the average cost of finding and producing a barrel of oil is about $5, and the most expensive oil currently being marketed in any serious quantity in the world today costs no more than $15 per barrel to produce.[3] The fact that this is true can be seen clearly in figure 4.1, which shows the actual (non-inflation-adjusted) sales price of a barrel from January 1984 to March 2007.[4]

It can be seen that as recently as 1999, the sales price of oil was as low as $11 per barrel, implying that production costs for the most expensive petroleum were less, and they certainly have not increased by more than 40 percent since.

As veteran commodities trader Raymond Learsy explains, "In normal commodities markets, sellers would start there, add the infrastructure cost and the expense of transporting and marketing the goods, and take a profit, which would be determined by competition with other producers. In a totally rational world, the world's cheapest oil would be used up first, and customers would turn to more costly sources when the least expensive ones dried up. In effect, this is what the Seven Sisters did when they controlled the market."[5]

Maybe so, but that is certainly not how it works today. To see the difference, take a look at figure 4.2. Here I have drawn reasonable guesses for the supply and demand curves for oil. (All supply and demand curves are guesses—no one really knows in advance exactly how much of a given product might sell at some other price than the current one.) The curve with the diamonds is supply; it goes up with price as more producers can do business profitably. The curve with the squares is demand; it falls with price, as less product can be sold when it costs too much. The place where they cross is where buyers and sellers agree, and thus is the free market price. In this case, with the assumed data, the answer comes out to 87 million barrels per day selling worldwide at a price of $20 per barrel. Most traders would agree that this is close to the right number for the free market price of oil today, and, in fact, it was the going price as recently as 2002.

However, functioning as a cartel, OPEC can survey the demand

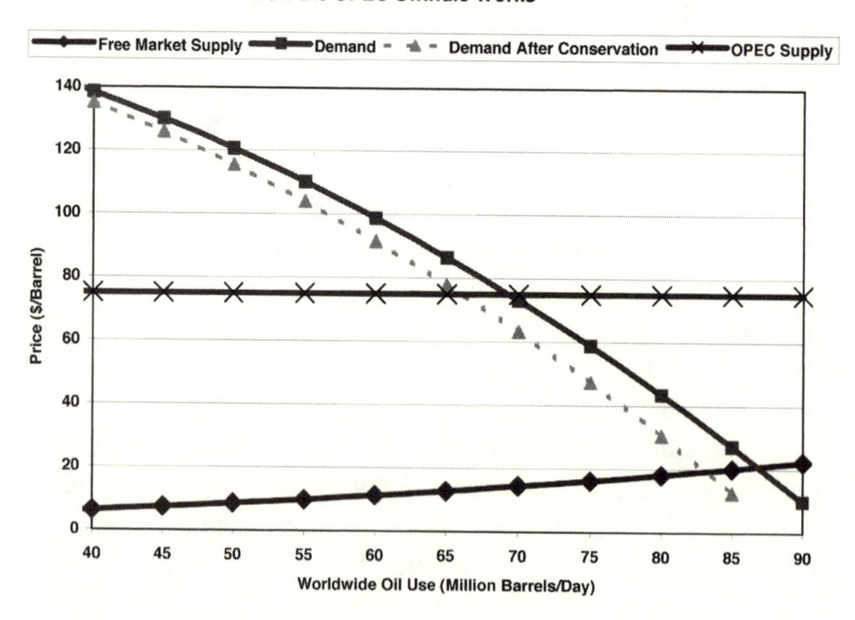

Figure 4.2. Supply and demand curves for oil worldwide.
Values shown are reasonable guesses selected for explanatory purposes, not necessarily corresponding to real data.

curve and simply decide where they want to intersect it. If they want a price of $75 per barrel, all they have to do is draw a line horizontally across the chart (as shown), and observe that the right answer (for them) is to limit worldwide production to 72 million barrels per day. As a result of such action, the world's oil-hungry customers will have to go without 15 million barrels of oil per day, but the OPEC members will make off like bandits. Consider, with an average production cost of $5 per barrel, at the $75 per barrel price they will make a profit of $70 per barrel, compared to the $15 per barrel profit they would reap at the free market price. True, their sales volume is a bit lower, but $70 per barrel profit multiplied by 72 million barrels per day works out to a profit of $5.04 billion *per day*, much better than the measly $1.305 billion per day profit the free market would provide.

Looking at figure 4.2, you can also see how weak conservation is as a strategy when deployed against OPEC. The dashed line with the triangle markings shows a new demand curve, moved somewhat to the left by a hypothetical successful global initiative to reduce consumption worldwide by 5 percent. In a free market, this would result in a slight decrease in the oil price. But all OPEC has to do is reset its quotas, and the price wouldn't change a bit.

In figures 4.3 and 4.4, you can see the actual non-OPEC and OPEC production of oil from 1973 to 2005, set against the inflation-adjusted oil price.[6] It will be observed that the non-OPEC production (shown in figure 4.3) follows a fairly smooth curve, but that the OPEC production (figure 4.4) varies wildly over very short time periods. These quick changes in the OPEC production are caused by the decisions of the OPEC ministers to make arbitrary adjustments in their quotas in order to drive up prices.

For example, look at the years from 1998 to 2005. Non-OPEC production rose slowly through this period. But between 1998 and 1999, OPEC dropped its production from 29.5 million barrels per day to 26.3 million barrels per day, a move that sent the price surging from $10 per barrel to $30 per barrel. Then, after jacking their production back up to 30 million barrels per day to take in the profit from the higher price

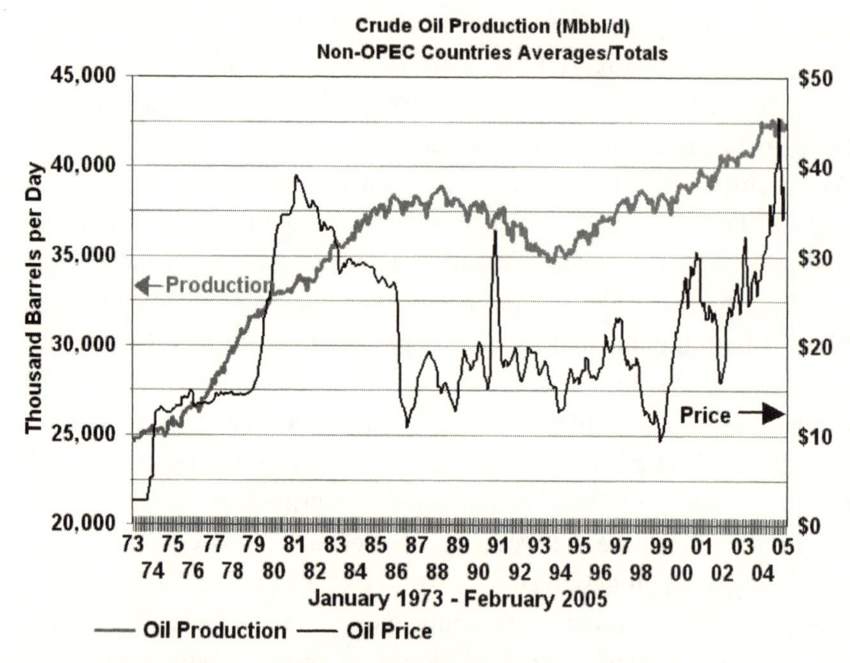

Figure 4.3. Non-OPEC production and inflation-adjusted (2004 dollars) prices for oil. *Source*: WTRG Economics, www.wtrg.com.

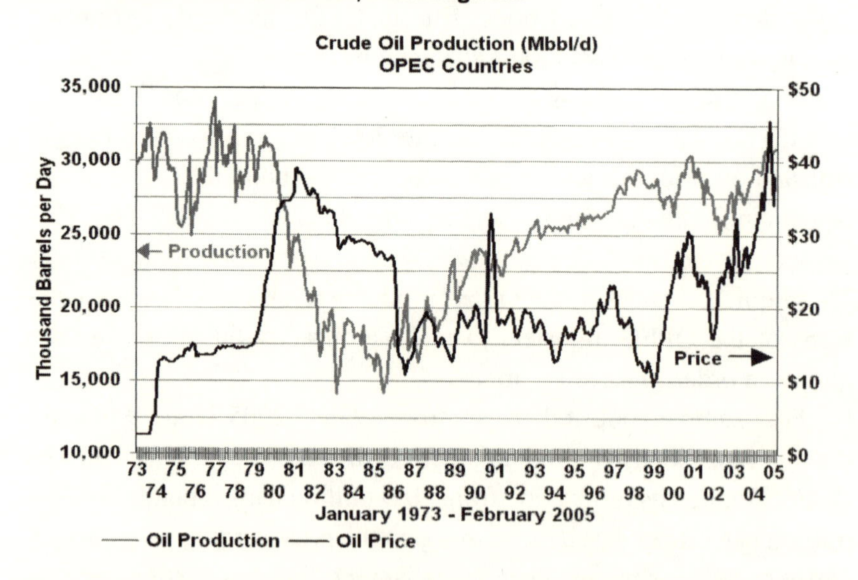

Figure 4.4. OPEC production and inflation-adjusted (2004 dollars) prices for oil. *Source*: WTRG Economics, www.wtrg.com.

(a move that caused prices to drop back to $15 per barrel in 2002), OPEC cut production again, all the way back down to 25 million barrels per day, causing the price to soar above $45 per barrel in early 2005 (and over $75 per barrel by mid-2006).

OPEC is effectively led and controlled by the Saudis. They have this control because they are the owners of not only the largest oil export capacity, but the cheapest. Saudi oil costs only $1.50 per barrel to produce. This fact gives them the whip hand over all the other OPEC members. If the other OPEC players try to cheat against their quotas (which they all like to do once the price has been run up), the Saudis have the ability to punish them by expanding production to crash the price. Since their oil is so cheap, the Saudis can still make plenty of money at $20 or even $10 per barrel, but most of the rest would be hit hard. Thus the Saudis enforce discipline upon the cartel. While grumbling occasionally, the rest willingly accept this discipline because they all know that it is necessary for the cartel to succeed in its mission of maximizing prices. Non-OPEC nations, including Mexico, Russia, Oman, and Norway, frequently choose to march in accord with the Saudi tune for the same reason. This expands OPEC power even more.

It should be noted that these actions by OPEC were, and are, illegal. Collusion by suppliers to fix prices is not only a crime under US law, it is banned by international law as well. The rules of the World Trade Organization (WTO) contain antitrust provisions that prohibit member nations from setting quota restrictions on import and exports. The WTO outlaws conspiracies to fix markets, and permits member nations to prosecute all parties to such conspiracies.[7] The US Justice Department would thus be entirely within its rights to initiate prosecutions against OPEC principals with interests in the United States (for example, Saudi royals), as well as against corporations, such as international oil companies, found to be acting in concert with OPEC. In addition, the imposition of retaliatory trade measures against OPEC nations would be fully justifiable.

The criminality of OPEC, however, is not just a legal matter. Consider the following: During 2005, as a result of OPEC price fixing, Saudi Arabia's high-priced oil business reaped that country's rulers more than $150 billion, to be spent on a combination of profligate waste, influence buying, and terrorism. Simultaneously, Kenya, a nation whose population of 36 million is half again as great as that of Saudi Arabia, managed to scrape up around $2.5 billion in export earnings,[8] and then had to use these precious funds to buy overpriced but badly needed fuel (from Saudi Arabia and other OPEC bandits), with only a pittance left to purchase farm machinery, replacement parts, and other vital equipment. Kenya, incidentally, is not even one of the world's fifty poorest nations.[9] Many others are much worse off.

The amount of money taken in by the Saudi kleptocrats in 2005 as a result of OPEC's illegal price fixing would be enough to *double* the foreign exchange of *sixty* countries like Kenya. Furthermore, as a result of the price fixing, the small amount of cash that the poor nations have can buy even less. And in 2006, the price of oil was nearly doubled yet again.

The injustice of this is appalling. It is one thing to be forced to pay $75 per barrel for oil when you make $20 per hour. It is quite another when you make $3 per day. OPEC's price fixing amounts to a huge, involuntary, viciously regressive tax on the entire world population.

Despite this, at least two groups of spokesmen have surfaced in the West to tell us that OPEC's banditry should be gratefully accepted, because it really serves the common good of all humankind. According to one school, OPEC is a blessing because the world is allegedly running out of oil, and by raising the price, the wise men of the cartel are helping us all to conserve. According to the other group, the world has unlimited oil, but the oil companies need a financial incentive to look for and develop it, so by dramatically raising the price, OPEC is doing its level best to make sure that there will always be plenty of oil for everyone.

Given the discussion in the previous chapter, I don't think it would surprise anyone to discover that many of the people who are saying these things are on the payroll of one part or another of the oil cartel. However, even though the financial links attached to many of these

OPEC apologists are rather clear, it is useful to deal with the arguments themselves.

Now it is true that raising the price of oil will tend to cut consumption, but not by much. Oil demand is very inelastic—it takes *enormous* price increases to effect any significant change in consumption. If we accept the demand curve hypothesized in figure 4.2, we see that a cut in oil consumption from 85 to 70 million barrels per day (an 18 percent reduction) needs a near *quadrupling* of the oil price to be enforced. The real historical data shown in figure 4.4 suggests that the situation is far worse: The quintupling of oil prices since 2001 was implemented by OPEC simply by cutting 5 million barrels—or 7 percent—out of the oil supply. During the same period, sales of low-gas-mileage cars such as SUVs continued to grow without missing a beat. If we actually wanted to enforce global petroleum conservation to any important extent through price increases, we would have to raise costs several hundred dollars per barrel. That would give the Saudis control of the world.

Fortunately, however, the claim that the world is running out of oil has no foundation whatsoever. Such claims have been made repeatedly in the past, and all have proven false. For example, in 1874 the state geologist of Pennsylvania, then the world's leading oil producer, estimated that the United States had enough oil for only another four years. In 1914 the Federal Bureau of Mines said we had only ten years of oil left. In 1940 the bureau revised its previous forecast and predicted that all our oil would be exhausted by 1954. In 1972 the prestigious Club of Rome, utilizing an inscrutable but allegedly infallible MIT computer oracle, handed down the ironclad prediction that the world's oil would run out by 1990.[10] The club said at that time that only 550 billion barrels were left to humanity. Since then we have used 600 billion barrels and are now looking at proven reserves of a trillion more.[11] Since 1972 there have been repeated predictions of imminent oil supply exhaustion published every few years by various authorities, and not one has come true. In fact, if we look at the ratio of proven reserves to consumption rate, the world has a bigger oil supply today than it *ever* has at any time in the past! The argument that we are threatened with near-term oil exhaustion is simply untrue.

While having no scientific data to support it, the "running out of oil" argument is nevertheless supported by Malthusian belief, which holds that since the world's resources are more or less fixed, population growth and living standards must be restricted or all of us will descend into bottomless misery.

As it conveniently provides a nice rationale for cutting the consumption of the most vulnerable, this ideology has historically been repeatedly used to justify various kinds of extreme exploitation, OPEC global looting being only the most recent example (others include social Darwinism, imperialism, and Nazism). However, as a scientific theory, Malthusianism is utterly bankrupt. All predictions based upon it have proven wrong, because human beings are not mere consumers of resources. Rather, we create resources by the development of new technologies that find use for them. The more people, the faster the rate of innovation. That is why (contrary to Malthus) as the world's population has increased, the standard of living has increased, and at an accelerating rate. The pompous Malthusians of the Club of Rome[12] had their infallible computer predictions of resource exhaustion proven wrong not just for oil, but *for every single commodity or mineral they discussed,* and all for the same reason: Humans are creative inventors. In the case of oil, we have invented a host of technologies for discovering and developing vast petroleum reserves that were simply impossible to find or recover in 1972. There is every reason to believe that this trend will continue, as most of the world, including nearly all of the seafloor, remains completely unexplored.[13]

This then brings us to the second argument, which tends to be advanced by American oil industry executives. After hearing your fill from Malthusian ideologues, these practical businessmen can sound absolutely refreshing. They say, "Yes, of course, we are not running out of oil. And yes, there certainly is plenty of world left to explore, but we need an incentive to do so. If you want us to keep finding the oil you want, you need to make it worth our while. So it's best that the price remains high." This is essentially the argument made by ARC Financial (an oil investment firm) economist Peter Tertzakian in his otherwise intelligent book, *A Thousand Barrels a Second.*[14]

The fallacy of this argument is that while it is no doubt true that a high current oil price will encourage increased exploration, the amount of such activity produced by the increased price is in no way commensurate with the magnitude of the global tax that it imposes. Consider this: At jacked-up prices of about $75 per barrel, the world's oil producers pulled in about *$2 trillion* during 2006, instead of the $500 billion or so they would have obtained in a free market. How much of the extra $1.5 trillion do you suppose they will put into oil exploration? No one knows for sure, but the correct answer is probably less than 2 percent. If our goal were increased oil exploration, we would do much better to set up a program, funded at, say, $30 billion per year, wherein the US government would offer competitive contracts paying private exploration companies to explore, creating forceful incentives for success with a 10 percent share in whatever reserves they discover. The government could then sell the claims so acquired to independent oil companies to develop, and use its 90 percent share to pay for the exploration program, plus probably a lot else. As for the other $1.47 trillion supposedly necessary to incite exploration, the consumers could keep it.

In other words, the OPEC-inflated oil price is simply a swindle. It does not benefit the world by preserving our precious petroleum fluids, and it is not necessary in order to make oil exploration occur. It is just a way to loot the world and radically expand the power of a group of cultists bent upon our destruction. There is no reason for us to tolerate it.

So, that said, how do we beat it?

WHY CONSERVATION WON'T WORK

When the subject of fighting OPEC comes up, the foremost idea that is generally advanced is conservation. Since OPEC is taxing us by selling us oil, and in fact using the demand generated by our purchases to help run up the price, we should simply use less. It sounds very sensible, yet it is completely unworkable.

The advocates of conservation break down into two groups. In the first instance, there are those who, for reason of their Malthusian convictions, favor conservation as a goal in itself, and are quite willing to support increases in the oil price in order to suppress consumption. These people, as we have already shown, are actually OPEC's allies. In fact, they have much more in common with the Islamists than simply economic goals, but we shall leave that subject for later.

In this section, however, I want to discuss the idea advanced by the second group, consisting of much more well-meaning people. Many of these people are quite hostile to OPEC. They understand the evil it is doing, either when it takes or when it gives, or both. They want to bring it down by boycotting its product. If I thought this idea had a prayer of working, I'd be for it. But it doesn't. It needs to be discredited because it proposes a strategy that guarantees defeat.

There are essentially only three ways to convince people to conserve: economic incentive, moral persuasion, or governmental action. None of them will succeed in this instance.

The most powerful persuader of the three is economics. Yet that is impractical here, because the objective is to *reduce* the price of oil. If the oil price is allowed to soar high enough to induce conservation, OPEC wins everything.

The notion of using moral exhortations to urge people to conserve is very nice, but as a practical strategy for reducing global oil consumption, let alone outmaneuvering OPEC, it is a joke. Whenever this idea is suggested, I am reminded of the Whip Inflation Now (WIN) campaign launched by the Ford administration during the mid-1970s to counter retail price hikes. (Whip Inflation Now! Rah, rah, rah!) Alternatively, one might think of it as being comparable to an attempt to alleviate world hunger through having church leaders call upon their congregants to eat less. A more serious approach is required.

That leaves the possibility of government mandates. These can certainly be forceful in their effect within the territorial jurisdiction of the United States, but for better or for worse, the United States is only one small part of the world. Demand for oil is a global issue. As I mentioned in chapter 1, and will discuss at greater length in the chapters to follow, there is a candidate mandate that could be enacted within the United States that *could* greatly affect the world energy market, but it is not one for conservation, which is what we are talking about here. Could any practicable US government conservation initiative lead to domestic consumption reductions large enough to influence the global oil price? Unfortunately, the answer is no.

Now admittedly, the term "practicable government initiative" used in the preceding paragraph has different meanings for different people. In November 2005, for example, I attended a World Technology Network workshop in San Francisco where energy conservation guru Amory Lovins argued that America could become oil-independent if its population were all induced to move to the vicinity of their workplaces. No doubt. But, thankfully, I don't think that the required Population Relocation Law would have a prayer in our political system. (Lovins, however, does have many good technical suggestions for reducing the energy consumption of particular items, such as buildings. If you are putting up a new factory and are concerned about the heating bill, look him up. It's his social ideas that go off the tracks.)[15]

To be realistic as an energy strategy, a plan must also be at least hypothetically, marginally, feasible politically. So utopianism of the Lovins sort is out. Instead we need to focus on policies that might actually be practicable. In the world of oil conservation ideas, the primary such candidate suggestion is that of increasing the standards for motor vehicle mileage.

Motor vehicle usage accounts for about half of all American oil use, so cutting it would certainly be a good place to start if one wanted to conserve oil. Moreover, there is some past success that can be pointed at in this respect. In 1976 the US government imposed the Corporate Average Fuel Efficiency (CAFE) standards on the auto industry. Those standards were met, and as a result, between that year and 1990, average American automobile fuel economy rose from 13 miles per gallon to 20 miles per gallon.[16] Despite this remarkable achievement, however, US gasoline consumption *increased* over the same period from 89 billion gallons per year to 103 billion gallons per year. But the engineering mileage improvements accomplished between 1975 and 1990 were the low-hanging fruit. Since 1990 there has been *no* further increase in vehicle mileage, and oil consumption has continued to rise, reaching 140 billion gallons per year in 2005.

It may be remarked that if not for CAFE, the rise in gasoline consumption would have been even worse, and that is true. But the point is, it didn't stop OPEC. Far from it. Despite CAFE and comparable or even more forceful measures introduced in many other countries, between 1975 and today world oil consumption rose from 50 million barrels per day to more than 70 million barrels per day.

Furthermore, regardless of the quadrupling of oil prices over the past six years, the rate of rise of oil consumption globally is accelerating, with increases over the past several years averaging about 1.7 million barrels per day each year. Only about 11 percent of this increase is occurring in the United States.[17] Most of it by far is occurring in China, India, eastern Europe, and Latin America. Everywhere the pattern previously seen is repeating: Once people become wealthy enough to buy a car, they do so. Auto sales in China doubled between 2001 and 2003, and have doubled again since. They are going to keep doubling.[18] In the United States today eight hundred out of every thousand people own cars. In China the number right now is only eight cars per thousand.[19] There are a *lot* more cars coming.

It would be harder to repeat the technical success of CAFE from 1975 through 1990 today, because the most obvious improvements have already been done. But let's say we could. This would imply raising average vehicle mileage from 20 miles per gallon to 31 miles per gallon. Then, over the next fifteen years, instead of US gasoline consumption rising to 190 billion gallons of gasoline per year (requiring an additional 5.7 million barrels of oil per day to produce), it would be expected to rise to only 162 billion gallons of gasoline per year (requiring an added 2.5 million barrels of oil per day to make). In other words, duplicating CAFE's achievement (whose impact, incidentally, was substantially augmented by a major US economic slowdown between 1974 and 1983) would not cut our oil consumption at all. Instead, it would reduce our expected rate of increase of oil usage by only 2.2 million barrels a day, during a period when the world as a whole is likely to raise its consumption another 30 million barrels per day. *Whatever demand we eliminate would be replaced fifteen times over.* Yet all OPEC has to do to win is to keep making trillions, which they can do even if worldwide demand merely holds at its present level!

In the fight against OPEC, the conservation strategy is a sure loser.

There was another area of energy policy where post-oil-shock US government policies did achieve some measure of success in reducing America's petroleum dependence. This was in the field of electric power generation. In 1974 about 17 percent of all US electricity was produced by burning oil and 18 percent from natural gas. However, as a result of the Carter administration's Fuel Use Act, which discouraged this, by 1985, oil-fired generators supplied only 4.1 percent of our electricity, and natural gas 12 percent. Today, though natural gas is back up to 18 percent, oil is down to 3 percent. The fraction of electricity that was once generated by oil has been taken up primarily by nuclear power, which went from 4 percent in 1973 to about 20 percent today.[20]

In view of the fact that nearly all of the Carter administration's numerous other energy initiatives were complete failures, the dramatic success of the Fuel Use Act is particularly remarkable. Why did it work so well? The answer is simple: The Fuel Use Act was not a fuel conservation policy. It was a fuel *substitution* policy. There was no reduction in electricity use during the period in question. Far from it—in 1978, when the act was passed, the United States generated 252 million kilowatts of electricity. By 1986, by which time petroleum had been substantially removed from the mix, we were generating 284 million kilowatts. In 2005, with oil cut down to provide just 3 percent of our electricity, we produced 443 million kilowatts overall.[21]

There are two points to be made here. First, this particular achievement cannot be repeated. With petroleum gone from our electric power base, we can't save significant additional quantities of oil by altering our electricity production or use. So, as aesthetically pleasing as they might be, windmills and solar panels are not the answer for fighting OPEC. (They can however, along with nuclear power, indirectly assist matters by making more natural gas available for methanol production. We shall discuss this later.) The second point, however, is much more important: *Conservation fails, but substitution works.* Indeed, while it takes a huge price hike to *reduce* energy use, it only takes a small edge to cause a *shift* from one energy source to another.

This is the key insight to grasp if we want to destroy OPEC.

To annihilate the oil cartel, we need to switch the world to a different fuel.

CHAPTER 5

CHANGING THE ENERGY TRUMP SUIT

In a number of card games, such as bridge and whist, the hand is frequently played with a special "trump" suit whose cards may overpower all others. For example, if spades are trump, then even the little two of spades will defeat the highest-ranked royalty of diamonds, hearts, or clubs. It is a critical element of strategy of such games to see to it that the trump suit selected is the one in which your own team holds most of the cards. If you fail to do this, and instead allow your opponents to elect their own long suit as the trump, you will end up playing the hand at a severe disadvantage.

The same is true in the energy game. Right now oil is trump, and in consequence the United States is playing the game with a losing hand. In table 5.1 I've laid out the facts that make this clear, by listing the world's leading oil producers, consumers, exporters, and importers. If all you look at is the production column at the left, the American position seems reasonably strong—we are the world's third-largest oil producer. But the more relevant information is given by the net import/export balances listed in the two columns at the right. When all is said and done, the United States is the rock-bottom worst loser in the world petroleum game. In fact, our losses every year are more

TABLE 5.1. THE WORLD'S LEADING OIL PLAYERS (MILLION BARRELS/DAY, 2005)[1]

Oil Production		Oil Consumption		Top Oil Exporters		Top Oil Importers	
Saudi Arabia	*11.1*	United States	20.7	*Saudi Arabia*	*9.1*	United States	12.4
Russia	9.5	China	6.9	Russia	6.7	Japan	5.2
United States	8.2	Japan	5.4	Norway	2.7	China	3.1
Iran	*4.2*	Russia	2.8	*Iran*	*2.6*	Germany	2.4
Mexico	3.8	Germany	2.6	*UAE*	*2.4*	South Korea	2.2
China	3.8	India	2.6	*Nigeria*	*2.3*	France	1.9
Canada	3.1	Canada	2.3	*Kuwait*	*2.3*	India	1.7
Norway	3.0	Brazil	2.2	*Venezuela*	*2.2*	Italy	1.6
Venezuela	*2.8*	South Korea	2.2	*Algeria*	*1.8*	Spain	1.6
UAE	*2.8*	Mexico	2.1	Mexico	1.7	Taiwan	1.0
Kuwait	*2.7*	France	2.0	*Libya*	*1.5*		
Nigeria	*2.6*	*Saudi Arabia*	*2.0*	*Iraq*	*1.3*		
Algeria	*2.1*			Angola	1.2		
Brazil	2.0			Kazakhstan	1.1		
				Qatar	*1.0*		

Note: OPEC members listed in *italics*.

than the next four biggest losers put together. For us, this game is a total disaster.

Now look who the winner is. In first place, Saudi Arabia, the primary bankroller of the worldwide Islamist movement.

A glance at the winners' column in table 5.1 will also tell you something else very important about the oil game. I've listed all the OPEC members in italics. Note how much bigger Saudi Arabia is than all the other OPEC members. The rest of the gang all export just 1 or 2 million barrels of oil a day each. The Saudis export more than *9*.

The Wahhabi terror patrons sell as much oil as the next four OPEC members combined. They are head and shoulders the leader of the group, and in fact, *without them, the gang would have no clear leader*. Hold that thought. We shall return it later.

The dismal American position in the global oil game is made even clearer if we consider the matter of petroleum reserves, which I place before you in table 5.2.

TABLE 5.2. GREATEST OIL RESERVES BY COUNTRY (2005)[2]

Country	Proved Reserves (billion barrels)	Reserves/Production (years)
Saudi Arabia	*261.9*	*64.6*
Iran	*125.8*	*82.1*
Iraq	*115.0*	*155.2*
Kuwait	*101.5*	*103.0*
UAE	*97.8*	*89.8*
Venezuela	*77.2*	*75.5*
Russia	60.0	17.3
Libya	*39.0*	*71.2*
Nigeria	*35.3*	*37.2*
USA	29.5	9.9

Note: OPEC members listed in *italics*.

Now it is quite true that new oil reserves are always being discovered, so despite the calculated reserve/production ratios presented in table 5.2, you should not expect the United States to run out of oil in 9.9 years. Nevertheless, the significance of the fact that the OPEC nations (shown in italics) have reserve/production ratios so much higher than we do cannot be missed. We will tap out first. In twenty or thirty years, they will be the only ones still holding any trump cards. If they are beating us badly in the oil game today, in the future they will simply trample us into the ground.

To see just how bad the situation is likely to become, take a look at figure 5.1, which shows the way the ownership of global oil reserves is projected to change over the coming years.[3] If matters are left on their current track, by 2020 the Islamic tyrannies will hold a position of overwhelming dominance with respect to the world's oil supply.

So if we want to win the energy game, we need to change the fuel trump suit. In cards, there are four different suits; the same is true in fuels. Only instead of spades, hearts, diamonds, or clubs, fuels can come as oil, coal, natural gas, or biomass.

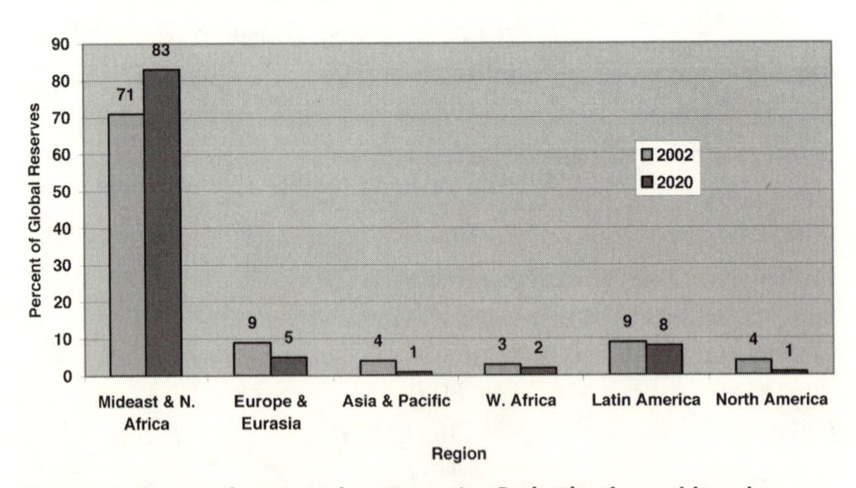

Figure 5.1. The looming strategic catastrophe. Projecting forward based on current data, by 2020, Middle Eastern regimes will hold 83 percent of global oil reserves. *Source*: "The Future of Oil," Institute for the Analysis of Global Security, http://www.iags.org.

Let's see how the fuel game changes if we switch the trump suit to one of these other possibilities.

In table 5.3, I've listed the balance of power if coal is trump. As you can see, the outcome in this case is very different. America wins this game, hands down.

Again, the reserve/production ratio should not be taken literally as time to exhaustion. More coal reserves are discovered whenever anyone bothers to explore, which is not that often, since known reserves are already so abundant. The key point is that if coal were the world's fuel trump suit instead of oil, America's role would change from pauper to powerhouse, while the Saudis and the rest of their ilk would go precisely the opposite way.

If instead the trump suit were natural gas, America still comes out fairly well. The numbers for this case can be seen in table 5.4, presented both in terms of billion cubic feet (bcf) and metric tons, or tonnes (1 metric ton = 1.1 English tons). I haven't shown figures for reserves, because depending upon what people decide to include, the reserve numbers range from frighteningly tight to astronomically vast. In contrast, the production statistics are at least knowable.

North America is currently roughly self-sufficient in natural gas. Europe, however, is a major importer, dependent upon Russia to make up most of its shortfall. Asia draws significant natural gas imports from the Middle East.

Finally there is the possibility of making fuel from biomass, derived from either agricultural crops or wild plants. In assessing biomass potential we are presented with a problem, in that good statistics exist only for commercial crops and forestry products that are actually harvested and sold. The capacity of various nations to produce such products is certainly indicative of their biomass production potential, but is hardly the whole story. For example, many countries could grow much more crops than they currently do, if there were a market for them. Furthermore, in transforming biomass into fuel, many parts of the plants that currently are not harvested, such as leaves, roots, and stems, could also be used. Finally, biomass for fuel can be produced

TABLE 5.3. WORLD COAL RESOURCES[4]

	Coal Reserves (2005) (billion tons)	Coal Production (million tons/year)	Reserves/Production (years)
USA	270.7	1,112	243
Other North America	8.8	85	104
Central and South America	21.9	74.9	292
Europe	63.8	806	79
Ukraine	37.6	69.3	543
Russia	173.1	308.9	560
Other Former USSR	39.8	115	346
Middle East	0.5	1.08	463
South Africa	53.7	267.7	201
Other Africa	1.8	6.18	291
Australia	86.5	391	221
India	101.9	443.7	230
China	126.2	2,156	59
Other Asia and Oceania	12.7	241	53
World Total	1,000.9	6,078	

TABLE 5.4. WORLD NATURAL GAS PRODUCTION (2004)[5]

Region	Gross Production bcf/yr	Gross Production (million tonnes)
North America	26,704	550
South America	4,541	94
Europe	11,890	245
Asia	10,637	219
Africa	1,019	21
Former USSR	28,157	580
Mideast (incl. N. Africa)	14,211	293
Australia & Oceania	1,462	30
Total	98,620	2,032

out of innumerable types of wild plants that are not commercial crops or forestry resources at all.

Currently, scientists estimate that about 220 billion tonnes of biomass grow each year worldwide.[6] About 4 percent of this, or 9.2 billion tonnes per year, are harvested by humans in the form of food crops and forestry products. At least another 6.2 billion tonnes of unused plant material are grown along with this commercial product but allowed to go to waste. In tables 5.5 and 5.6, you can see where all this stuff is now grown. Not surprisingly, the biomass trump-suit heavy hitters are Asia, North America, Europe, and South America, but for the first time, Africa has a reasonably good hand, too. I mention this as a strong positive point relating to the possibility of a biomass-based fuel economy, because if we are ever to achieve a decent and just world, everyone needs to have a piece of the action.

Now that's a lot of statistics to wade through, and it may be a bit confusing. In particular, readers may be wondering how a tonne of biomass compares to a barrel of oil or a billion cubic feet of natural gas. Oil, coal, gas, and biomass all vary somewhat in energy quality, but if you lined them up against each other on the basis of weight, a

TABLE 5.5. WORLD AGRICULTURAL PRODUCTION (MILLION TONNES, 2003)[7]

Region	Food Crops	Forestry Products	Total
North America	584	1,469	2,053
South America	256	602	858
Europe	451	848	1,299
Asia	1,142	1,415	2,557
Africa	176	685	1,037
Former USSR	174	331	505
Mideast (incl. N. Africa)	219	76	295
Australia & Oceania	52	100	152
World Total	3,651	5,594	9,245

TABLE 5.6. WORLD BIOMASS PRODUCTION FROM RESIDUE (MILLION TONNES 1987)[8]

Area	Agricultural Waste	Forestry Waste	Total
USA & Canada	561	582	1,143
Europe	492	301	793
Japan	39	28	67
Australia & NZ	126	26	152
Former USSR	394	306	700
Latin America	825	243	1,068
Africa	465	276	741
China	793	177	970
Other Asia	1,443	470	1,913
Total	5,137	2,413	7,550

TABLE 5.7. WORLD MATERIAL RESOURCES FOR FUEL PRODUCTION

Fuel Resource	Billion tonnes/ year	Billion tonnes/year/ Oil Equivalent
Oil	3.39	3.39
Coal	5.53	3.87
Natural Gas	2.03	2.64
Farm & Forestry Products	9.25	3.24
Farm & Forestry Residues	7.55	2.65
Total Biomass Production	220	77

good rule of thumb is that 1 gram (gm) of biomass = 3.5 kilocalories (kcal), 1 gm coal = 7 kcal, 1 gm oil = 10 kcal, and 1 gm natural gas = 13 kcal.[9] There are about 7.33 barrels of oil per metric tonne, and 1 million cubic feet of natural gas weighs about 20.6 tonnes. Pulling all this together, the comparative lineup of possible fuel materials for the world is laid out in table 5.7.

Examining table 5.7, you can see that the energy value of the world's oil product is only about one-fifth the combined energy of the world's annual fuel material yield. In round numbers, it can be seen that each of the four fuel suits—oil, coal, gas, and biomass—offers comparable energy resources. Unfortunately, however, the trump suit is oil, and this is the one where we are weakest and the enemy is strongest. Oil is trump because, while all four suits can provide raw energy, only oil can currently provide energy in the liquid form necessary for powering practical motor vehicles and aircraft, and most desirable for driving ships and trains as well.

But what if the rules were changed so that oil was no longer trump? What if, as sometimes happens in bridge, the energy game were played with no trump suit, and all its four suits put on an equal footing? In that case, all of the four could be called upon to produce liquid transportation fuels. That sounds promising, but all of the commercial product of the existing coal, gas, and biomass sectors is already spoken for. To get net liquid fuel out of them, we would have to expand their production. Could we do that on the scale required to make a real difference? In war, as General Douglas MacArthur famously put it, there is no substitute for victory. Our goal needs to be to destroy OPEC, not just reduce its looting a bit at the margins. How much new fuel would we have to draw from our three strong suits in order to sink the oil cartel? Let's see.

While OPEC produces about 40 percent of the world's oil, it exports only about a third. Furthermore, of the 24 million barrels per day that OPEC exports in total (see table 5.1), the Saudis account for 9.1 million, which is roughly 12 percent of worldwide oil production. If we could replace those 9.1 million barrels with usable nonpetroleum liquid fuel, we could dictate terms to Saudi Arabia.

For example, one attractive option would to be to inform the Saudis that, in view of their past and ongoing misconduct (including oil price fixing, promotion of terrorism, damages to the World Trade Center, etc.), in the future, the whole of their oil export product would have to be sold to the United States government at $10 per barrel. They could accept these terms—or sit on their oil and go broke. Since we would have the required replacement fuel in hand, shutting down production would provide them with little leverage, so ultimately they would be forced to capitulate. At $10 per barrel, the Saudis could still make money selling oil, but they would be forced to expand production volume to the max. Having obtained the full, enlarged Saudi export product at $10 per barrel, the US government could then dump it on the world market at $30 per barrel. This would stabilize global oil prices at that level, save consumers worldwide trillions of dollars, and, incidentally, put about $200 million per day ($73 billion per year) into the US Treasury.

Such a move would decapitate OPEC and effectively bankrupt the prime promoters and paymasters of global terrorism. Throw in an extra 2.6 million barrels per day of substitute fuel, and we can shut down Iran as well. That would mean no money for Hezbollah or for the mad mullah's nuclear bomb program. Between those two accomplishments, most of our present strategic problems would be solved, so let's take that as our baseline plan. We need 11.7 million barrels per day of replacement fuel to pull it off. That comes out to one-sixth of the total world oil supply, or 0.54 billion tonnes per year.

Very well, so let's go back and take another look at table 5.7. Between coal, natural gas, and farm and forest products and residues, we are currently producing the energy equivalent of 12.3 billion tonnes of oil per year. To make up an oil shortfall of 0.54 billion tonnes (or less, since only 57 percent of each barrel of oil is actually used to make motor vehicle fuel) we would have to increase production across these other categories by a mere 4.4 percent.

This certainly could be done. The only issue is whether we can put the power game on no-trump rules by actually converting coal, natural gas, and biomass into their energy equivalent in usable liquid fuel. In point of fact, we can. No Manhattan Project will be required to discover how. The chemical knowledge required to do it is already quite well established, being hundreds—and in some cases thousands—of years old. All we need to do is make alcohol.

MAKING ALCOHOL FUEL

There are many different kinds of alcohols, and all may be used as fuels. However, for our purposes here, the two most important alcohols are the two simplest: methanol (CH_3OH) and ethanol (C_2H_5OH).

Ethanol is the kind of alcohol that we drink in alcoholic beverages. It has been made since prehistoric times by using microorganisms to ferment the sugar (and starch, which is a biological precursor to sugar) content of plant material. All food crops contain sugar and starch, which are the primary sources of their caloric value. Acting on this insight, humans over the millennia have made alcohol by fermenting grapes, honey, wheat, barley, rice, sugar, molasses, apples, pears, potatoes, corn, and many other crops, each into its own characteristic drink. Basically, if it's a plant, and you can eat it, you can make booze out of it. The sweeter it is, the more ethanol you will get. This is why using sugarcane is the most economical way to produce ethanol, but almost any crop with a high caloric value will do. Thus, in recent years in the United States, a major ethanol industry has been created based on the conversion of corn. Corn, however, also contains much protein and other nutritional components not utilized in ethanol production. In most commercial corn-to-ethanol plants in America today, these components are saved and processed for sale as high-value animal feed.

That said, the food crops used as a basis for ethanol production through this technique have significant commercial value, and that puts a floor under the production cost of fermentation-based ethanol. For sugar, that cost is about $1 per gallon, while for corn it is around $1.50 per gallon (without any subsidy). Ethanol has about two-thirds the energy value per gallon as gasoline, so these prices correspond to gasoline sold at $1.50 and $2.25 per gallon, respectively. These (before-tax—most gasoline sold in the United States is taxed about $0.50 per gallon) gasoline prices, in turn, correspond to oil priced at $36 and $54 per barrel. So long as oil is pegged above this level (as it currently is), crop-fermentation ethanol can beat the price of gasoline. Should the price of oil drop below such levels, some combination of tariffs, subsidies, or preferential taxes would be required to keep crop-fermentation ethanol competitive.

Only limited fractions of the plants that farmers grow actually are sold as commercial crops. Large quantities of vegetative material, such as the stems, roots, and leaves of corn plants, for example, are left to waste. This is also true of the leaves of trees that are discarded in the fall, and vast quantities of grasses and weeds that grow, yellow, and fade every year without being eaten or otherwise used by any living creature. If this material could be turned into ethanol, its price and availability would improve radically. All such material, however, is primarily cellulose, which cannot be readily fermented into ethanol. For this reason, considerable research is currently under way in a number of places worldwide to try to discover or engineer some microorganism that can transform cellulose into a starch or a sugar, which would then be fermentable. In principle this should be possible, because grazing animals such as horses, deer, and cattle perform exactly this trick in their stomachs all the time—this is why they can eat leaves. However, isolating the bacteria or their enzymes that perform this chemistry for the grazers and getting them to serve humanity in the same capacity in an industrial setting has proven to be a challenge. Eventual success for cellulosic ethanol technology appears highly probable, but we don't have it yet. Fortunately, however, there are simpler techniques that can be used to make usable alcohol fuel out of biomass, and much else. This brings us to the subject of methanol.

Methanol, commonly known as "wood alcohol," is the simplest liquid fuel molecule known to chemistry. As such it can be readily manufactured out of virtually any kind of organic material, including not only *every* kind of biomass, whether edible or not, but coal, natural gas, human and animal metabolic wastes, and municipal trash. Since its potential sources are so vast, varied, and cheap, methanol promises to be an inexpensive fuel—and in fact, it is. During 2007 the wholesale price of methanol, manufactured and sold without a subsidy, was $0.93 per gallon. Methanol has about 54 percent the energy density of gasoline, so this price is the equivalent of gasoline selling for $1.72 per gallon, before taxes.

Depending upon the source material, there are a number of different ways to make methanol, but they all come down to the same few chemical reactions. Let's start with coal.

Coal is basically carbon (elemental formula C), with some impurities mixed in. Carbon can be reacted with steam (H_2O) to form carbon monoxide (CO) and hydrogen (H_2), which come off as a mixture that has traditionally been called synthesis gas, or "syngas," because it can be used to synthesize many chemicals, including methanol, as we shall discuss shortly. Performing such steam reformation of coal in this way requires an energy input of +40 kcal for every mole unit reacted. (A mole is the amount of a chemical equal to its molecular weight in grams. For example, carbon has molecular weight 12, so a mole of carbon would be 12 grams.) This energy needs to come from somewhere, and in this case an obvious way to get it would be to burn some additional coal in air.

The combustion of coal provides an energy output of 92 kcal/mole, more than twice as much as is needed to drive steam reformation. In other words, by burning one unit of coal, we get enough energy to turn a bit more than two other coal units into syngas.

Once we have syngas, we can change the proportions of carbon monoxide and hydrogen within it at will by performing a reaction known as the water gas shift. This works by combining carbon monoxide with water to produce carbon dioxide and hydrogen. The water gas shift is close to energetically neutral. That's fine, though, because we are running it to change our syngas H_2/CO mixture ratio, not to produce power. The steam reformation of coal produces syngas with hydrogen and carbon monoxide in equal proportions, whereas to make methanol, we want our syngas to have twice as many hydrogen molecules as carbon monoxide molecules. By running the water gas shift reaction, we make more hydrogen while getting rid of some carbon monoxide, which is exactly what we need to do to get the syngas mixture we want. Once this is done, we react the hydrogen and carbon monoxide together to make the methanol (CH_3OH).

If you are new to chemistry, the preceding discussion may sound like an involved process, but to chemical engineers, it's all very straightforward. The methanol synthesis reaction can be done over a copper catalyst in a simple steel reactor vessel operating at a temperature of 250°C and a pressure of twenty atmospheres. These are easy conditions to create, and in fact the entire process I've just described is gaslight-era technology.

We can use similar reactions to turn natural gas into methanol. Natural gas is mostly methane (CH_4), and it can be reacted with steam to produce syngas just like coal can. The steam reformation of methane requires +59 kcal/mole, but plenty of energy to drive it can be provided by burning a bit of methane, which yields 205 kcal/mole.

Alternatively, one can transform methane directly into syngas by reacting it with just enough oxygen to partially oxidize the methane into carbon monoxide.

Either way, once we have the syngas, the process to make the

TABLE 5.8. CHEMISTRY TOOLKIT:
USEFUL REACTIONS FOR PRODUCING METHANOL

Fuel	Equation	Energy input	Name of Reaction	Product
Coal	$C + H_2O \rightarrow CO + H_2$	+40 kcal/mole	Steam Reformation	Syngas
Methane	$CH_4 + H_2O \rightarrow CO + 3H_2$	+59 kcal/mole	Steam Reformation	Syngas
Methane	$CH_4 + 1/2O_2 \rightarrow CO + 2H_2$	-7 kcal/mole	Partial Oxidation	Syngas
Biomass	$C_4H_6O_3 + H_2O \rightarrow 4CO + 4H_2$	+ Varies	Steam Reformation	Syngas
Carbon Monoxide	$CO + H_2O \rightarrow CO_2 + H_2$	~0 kcal/mole	Water Gas Shift	Hydrogen
Coal	$C + O_2 \rightarrow CO_2$	-92 kcal/mole	Combustion	Energy
Methane	$CH_4 + 2O_2 \rightarrow CO_2 + 2H_2O$	-205 kcal/mole	Combustion	Energy
Syngas	$CO + 2H_2 \rightarrow CH_3OH$	-23 kcal/mole	Methanol Synthesis	Methanol

methanol proceeds just as it would with coal. Again, this is all strictly tried-and-true nineteenth-century chemical engineering.

A more modern way would be to use a special catalyst to oxidize methane directly into methanol. This saves a process step and thereby cuts costs. A number of such direct oxidation methanol plants are currently under construction.

Biomass, which is primarily cellulose, can also be used to make methanol. Cellulose is a complex long-chain molecule, with an approximate formula of $C_4H_6O_3$, with small amounts of nitrogen, traces of sulfur, and a few other odds and ends mixed in. Again, the strategy is simply to turn it into synthesis gas by reaction with steam, using heat provided by burning a minority of the material to drive the conversion of the rest.

Once we have the syngas, the rest of the process follows the same path as that needed to convert coal or natural gas. And the beauty of it is that *any* plant material, without exception—from weeds and fallen leaves to swamp cattails and the vast floating growths that clog innumerable rivers in Latin America and Africa—can be used as feedstock for the process.

As for converting trash, it doesn't matter whether the feedstock is composed of packaging materials, old rags, used candy wrappers, plastic forks, or Styrofoam coffee cups. The stuff is all just compounds of carbon, hydrogen, and oxygen, with a few impurities thrown in here and there, and all of it can be pyrolyzed and reacted with steam to produce synthesis gas, and then methanol.

The chemical equations specifying all the reactions we've discussed are shown in table 5.8. In table 5.8, a positive energy input means that energy is needed to drive the reaction, while a negative energy input means that the reaction produces energy.

The chemistry needed to dethrone oil from its trump-suit status is well understood. We can readily convert our fuel strong suits into an alcohol supply bountiful enough to wash OPEC off the map.

The only issue is that we need to have cars and trucks that can use it. This will be the subject of our next chapter.

CHAPTER 6

THE TECHNOLOGY TO BREAK OPEC'S CHAINS

There is no Nobel Prize for engineering. The closest thing to it is the Charles Stark Draper Prize awarded by the US National Academy of Engineering. While not as famous as the Nobel, the Draper is still rather prestigious, and comes with a $500,000 check. Those so honored have included the inventors of the turbojet, the integrated circuit, and the Internet—and collectively represent a roll call of the creators of the modern technological age. No woman has ever won it. There is, however, one who should have.

Her name was Roberta Nichols. Born in 1931, her dad was an engineer at Douglas Aircraft, and, despite the near total exclusion of girls from engineering schools in her day, she saw no reason why she couldn't become one as well. Working with her father, she repaired vintage automobiles, and built and piloted racing cars and boats. During one race on the Bonneville Salt Flats, she took her souped-up 1929 Ford Model A to over 190 mph. On another occasion, she raced her drag boat, the *Witch* (whose engine she had converted to run on methanol), to 131 mph, setting a world speed record that stood for three years. "I just grew up not knowing that girls weren't supposed to like to do those kinds of things," she said, many years later. So she became an engineer, and a very good one, too. Her passion was the internal combustion engine, and after a pause to have and raise children, her career advanced rapidly. She broke the glass ceiling. By the late 1970s, she was leading the research in alternative fuel vehicles at Ford.

The time was certainly right for a talented inventor of alternative fuel cars. The nation had received a massive blow from the first Arab oil embargo, and was about to receive a second shock following the Islamist revolution in Iran. During the 1970s, numerous committees of the Department of Energy, the National Academies of Science and Engineering, the Environmental Protection Agency, and other august advisory bodies had met, conferred, and issued reports, all converging on the view that America needed an alternative to foreign oil—and that methanol could well be the best answer. The environmental movement also was on the rise, adding new force to the issue of air pollution, particularly in Southern California, where this always palpable problem was becoming ever more urgent. Again, methanol vehicles appeared to be a solution.

Nichols had grown up in Los Angeles, graduating with a bachelor's degree in physics from University of California at Los Angeles, and then obtained advanced degrees in environmental engineering from University of Southern California. She had played a formative role in launching the California Energy Commission (CEC), and had many friends among the environmentalists who ran it after her departure for Ford. Working together with these contacts, she seized the moment and sold the state government on the idea of launching a major program to prove the practicality of methanol-powered automobiles. Then she went back to Ford and fought an internal campaign to get the company to partner on the project. To prove that methanol cars would work, she singlehandedly converted a Ford Pinto to methanol. Overcoming significant bureaucratic opposition, she got her way.

The experiment began in 1980, with Ford supplying twelve specially designed methanol-fueled Pintos to the state, then it picked up steam over the next several years. By 1983 California had more than 600 methanol-fueled cars operating, including an impressive fleet of 561 Ford Escorts.

The methanol vehicles were a great technical success. Methanol is 105-octane fuel, and its use in pure form in the California state fleet cars increased their effective horsepower by 20 percent. As Nichols recalled in one of her last papers, "[T]he drivers loved the performance."[1] With a compression ratio of 11.8:1, fuel efficiency was increased by 15 percent. The cars were calibrated to achieve an advanced emission standard, including 0.4 nitrogen oxide (NOx), which at that time gasoline-powered cars were unable to meet. The methanol cars met that standard when new, and not only that, they still met it with no deterioration whatsoever after 50,000 miles of use. Overall, the methanol fleet racked up some 35 million miles of real-world travel, and came through with flying colors. Indeed, by 1990, after seven years of use and abuse, more than 90 percent of the original methanol Escorts were still running strong.

There was, however, a problem. Methanol only contains about half the energy per gallon of gasoline, so even though the methanol engines were 15 percent more efficient in utilizing this energy, distance traveled per gallon was only about 57 percent of that attainable with gasoline. Ford tried to compensate for this by increasing the size of the fuel tank, but there were limits to how much capacity they could add given the fact that they were retrofitting an existing gasoline automobile model. As a result, the methanol Escorts could travel only about 230 miles before refueling. Such a range limit would only be a minor inconvenience for a gasoline vehicle, because there are gasoline stations everywhere. But California had only established twenty-two methanol refueling stations statewide. As Nichols put it: "It was clear that this number of stations was totally inadequate for the drivers of these vehicles to feel comfortable. They had to constantly monitor the fuel gauge and carefully plan their routes."[2] Furthermore, with only six hundred methanol cars operating in a state with more than 10 million gas-fueled vehicles, a private gas station owner would have to be insane to waste one of his pumps on the methanol market. In short, there weren't enough methanol stations to induce anyone other than the California Energy Commission to buy a methanol car, and until and unless there were a million or more methanol vehicles on the road, that would remain the case indefinitely. This chicken-and-egg problem was a showstopper. The methanol car might be a complete technical success, but unless something was done, it would also be a technological dead end.

In retrospect, the need for operational infrastructure to support the methanol vehicles may seem obvious, but engineers like to think about how to make machines work, and the environmentalists wanted to see a car that wouldn't pollute, so for the CEC/Ford team achieving technical success with the cars themselves must have seemed like the first priority. But having accomplished that, they were shoved face to face with the brutal reality that for a new technology to make its way into the world, intrinsic merit is not enough. In engineering, as in politics, nothing is so difficult or so perilous as the creation of a new order, and at birth, an infant invention that hopes to do so is confronted by a technical and economic reality that has been shaped not by its own needs, but by those it aims to usurp. Before the new technology can become strong enough to make the world of its desires, it must live and grow to power within the world as it is. The methanol cars might be a driver's delight, but unless they could function in a world *without* methanol stations, they would forever remain a footnote: just one more curious hobbyhorse technical demonstration instigated and toyed with for a little while by eccentric bureaucrats and playful engineers with other peoples' money to spend and nothing better to do with their time. That was not the fate Nichols intended for her project.

There was only one answer. If they were ever to make their way into the economy in numbers big enough to make a real difference to either energy independence or the environment, the methanol cars would also have to be able to run on gasoline.

Now, the technical issues associated with building methanol-only cars are not particularly great. Methanol is more corrosive than gasoline, so superior-quality materials need to be used in the fuel line. A methanol engine burns with more fuel and less air than gasoline, and so the fuel and air inputs for the engine need to be set accordingly.

Methanol is a bit harder to ignite than gasoline, but burns cooler and cleaner, and its high octane offers the opportunity to achieve an increased compression ratio, higher fuel efficiency, and better vehicle performance. These considerations variously complicate and ease the designer's task, but at the end of the day, an engineering team attempting to develop a methanol-only automobile is faced with a technical challenge roughly equivalent to that involved in creating a good gasoline-driven vehicle. In fact, in the 1960s many drivers, including Nichols, had independently converted their racing cars to methanol, preferring it for its safety advantages and higher octane. As we have seen, Nichols's alternate-fuel vehicle team at Ford had no difficulty solving such straightforward matters for heavy-use commuter cars as well, and the all-methanol vehicles they shipped to the CEC program during the early 1980s worked just fine.

Creating a mixed-fuel car, however, is another question altogether. Again, it would not be a particularly difficult challenge if the designer could be informed in advance that the car would run on some specified mixture, say, 40 percent methanol, 60 percent gasoline (M40), or whatever. But for a methanol/gasoline car to do what the Nichols team now realized it really needed to do, it would have to be able to run not on one specified fuel mixture, but on *any* arbitrary mixture. A car might start out the day with a full tank of methanol (M100), and then fill up with gasoline when the tank was three-quarters empty. From that point on, it would be running on M25, until such time as it refueled again, at which time its fuel mixture would change unpredictably to something else. The challenge was baffling. *How can one possibly design an automobile engine to work well without knowing what fuel it is going to use?* The only way to do it would be to have an engine that could actively change its behavior in immediate response to the quality of its fuel. But how?

An opening was provided by the Dutch inventor G. A. Schwippert, who in 1984 patented an optical sensor that could determine the alcohol content of a methanol/gasoline mixture by measuring the fluid's index of refraction (light-bending properties). Making use of this device and the new technology of electronic fuel injection then coming into general use, Nichols and her Ford engineering team collaborators Richard Wineland and Eric Clinton devised a scheme whereby a Schwippert sensor would be used to assess the alcohol content of the fuel in real time as it was being fed to the engine. The computer that controlled the car's electronic fuel injector (EFI) would then interpolate proportionately between the desired air/fuel ratios of pure methanol and pure gasoline to determine the correct air/fuel ratio for the mixture of the moment. This done, the computer would give the EFI its instructions for the right amount of fuel to feed to the engine. No matter what the fuel mixture might be, the EFI would always know how much to pump to make the engine operate correctly.

This design concept was laid out by Nichols, Wineland, and Clinton in the historic US patent 4,706,629, "Control system for engine operation using two fuels of different volumetric energy content,"[3] filed in February 1986. Together with their two other patented inventions covering the issues of spark ignition timing[4] and differential fuel volatility,[5] it represented the first complete practical system to enable an automobile to run omnivorously on any mixture of alcohol and gasoline.

It was a breakthrough. Nichols, Wineland, and Clinton had invented the flex-fuel car.

Nichols lost no time in putting her team's invention into practice. In 1986, even before the ink on the patent applications was really dry, she rushed an experimental methanol/gasoline flex-fuel Ford Escort to the California Energy Commission for field testing. The next year she followed with 7 methanol/ethanol/gasoline flex-fuel Ford Crown Victorias, and in 1989 she sent 183 more. This was just the beginning. The flex-fuel cars were so successful that in the early 1990s Nichols was able to get Ford to launch a full-production run, and some 8,000 methanol/gasoline flex-fuel Ford Tauruses were shipped to the state. Most significantly, for the first time, many of the cars were bought by the general public.

Realizing that something important was going on, the other auto manufacturers finally roused themselves and created their own flex-fuel concepts. By the end of the 1990s, General Motors had shipped the CEC 1,512 methanol/gasoline flex-fuel vehicles; Chrysler, 4,730; Volkswagen, 53; Nissan, 17; Toyota, 8; and Mercedes-Benz, 5.

The cars worked well. As the CEC's Tom MacDonald reported in a summary paper on the program published in 2000,[6] the more than fourteen thousand methanol/gasoline flex-fuel vehicles involved demonstrated "seamless vehicle operation on methanol, gasoline, and all combination of these fuels." Other conclusions from the CEC's massive field trial included "FFV engine durability can be expected to match that of standard gasoline vehicles. . . . An incremental improvement in FFV emissions was observed. . . . An incremental fuel efficiency is achieved using methanol. . . . Health and safety related issues that had undergone long examination and debate with respect to methanol proved largely insignificant in the expanded FFV fleet

Figure 6.1. Roberta Nichols fueling one of the first flex-fuel cars, circa 1986. Photo courtesy Ford/Peter Arnold, Inc.

demonstrations, with few, if any, reported incidents attributable to methanol use."

The path to a new world had been opened.

Times change and priorities alter. During the 1990s, with the price of oil hovering at low levels, both the nation and the CEC lost focus on the goal of energy independence, and, with it, on methanol. Interest in flex-fuel cars, however, was maintained by the farm lobby, which saw in them a means to expand ethanol sales. Thus, for the past decade or so, flex-fuel vehicles have been designed primarily for ethanol use. The technology has changed as well. Instead of using the early Nichols-Wineland-Clinton FFV approach of sensing the fuel content prior to injection, nearly all flex-fuel cars today use a method based on sensing the content of the exhaust. This is a better approach because it is not only cheaper, but more versatile. An optical sensor can't really tell which alcohol is mixed with the gasoline, and so it would have a hard time dealing with a situation where the type of alcohol used is unknown. In contrast, a system based on a sensor that sniffs the car's exhaust doesn't need to know what the fuel is at all. It simply assesses whether the engine is running lean or rich, and instructs the EFI to add fuel or air accordingly. Thus, while the current production models of FFVs are certified only for ethanol/gasoline mixtures, they can actually run on methanol/gasoline or methanol/ethanol/gasoline mixes as well. Because of the simplified system, there is almost no price differential between flex-fuel cars and their gasoline-only counterparts. As of this writing, about 6 million have been sold. That's still not enough to make any difference for energy independence, but it's enough to show the way.

Roberta Nichols died of leukemia in 2005, her dream still unachieved, but perhaps, for someone with vision as far-seeing as hers, finally in sight. Looking at it in many ways, one cannot help but admire her life and her accomplishment. Personally, however, I find especially delightful the historical irony that fundamentalist Islam, which detests uppity women, nonconformists, innovators, and liquor, should ultimately face its comeuppance from the brainchild of a gutsy, mold-breaking inventress with unshakeable faith in the power of alcohol.

ALCOHOL FUELS AND THE ENVIRONMENT

Roberta Nichols was dedicated to the goal of a clean environment, and so were her friends in the CEC who enabled the large-scale demonstration project that put methanol and flex-fuel cars on the map. I mention this because many in the environmental community today tend to be reactively suspicious, or even axiomatically hostile, to new technologies. It would be both ironic and extremely unfortunate if they, not knowing this history, were to adopt such a stance toward the widespread adoption of methanol fuel today. Methanol-powered cars were an environmentalist baby, advocated by activists in the face of then-existing economic disincentives for their use. There were good and substantial reasons for that advocacy, and they remain valid today. Restating them is extremely important. After all, with methanol producible without a subsidy for as low as $0.93 per gallon (its mid-2007 wholesale price), and gasoline running near $3.00 per gallon, the economic case for switching to methanol now speaks for itself. The technological feasibility of alcohol flex-fuel cars has been proven, and the existence of an enormous worldwide resource base to sustain them is apparent. If the environmental case can be nailed down, there should be nothing to stop us from moving forward.

In the 1970s and 1980s, when the environmental argument for methanol conversion was first made, it centered upon the superior potential of alcohol fuels for mitigating the immediately pressing problems of air pollution and toxic spills, both on land and on water. To these classic environmental issues, more recent times have added a new issue, that of coping with the longer-term problem of global warming. In this section we will focus primarily on traditional environmental concerns, leaving most of our discussion of countering the greenhouse effect until chapter 10. Suffice at this point to say, however, that the long-term need to address the issue of global warming strengthens the already rock-hard environmental case for alcohol fuels even further.

A fundamental difference between alcohol fuels and petroleum fuels is that alcohols can mix with water, whereas oil, gasoline, kerosene, and virtually all other petroleum products cannot. The earth is a water world, covered by a huge and extremely active hydrosphere. The fact that alcohols can dissolve in this ocean, and that alcohols are readily consumed by common bacteria, means that long-term environmental degradation caused by uncontrolled releases of alcohols is *impossible.* Today, a quarter century after the *Exxon Valdez* oil tanker disaster devastated twelve hundred miles of coastline, thousands of sea otters are still being killed by eating polluted clams. If, however, the *Exxon Valdez* had been carrying alcohol instead of petroleum when it wrecked, the threat to wildlife would have been rendered harmless within hours, or days at most, and the past occurrence of the event would have been made undetectable within months. Instead of hanging about for decades as a noxious oil slick, the alcohol cargo would have simply washed away and been diluted to nothing in the vastness of the sea. These same considerations hold with respect to possible seepage of methanol or ethanol into groundwater from defective pumping stations, crashed or abandoned automobiles, wrecked tanker trucks, leaky lawnmowers, or any other land-based source. On many lakes frequented by recreational powerboats today, an iridescent petroleum scum dangerous to wildlife and obnoxious to swimmers can be widely observed. If those powerboats were running on alcohol, that pollution would not exist.

Of the two alcohol fuels under consideration, ethanol is actually edible, while methanol is toxic—but then, so is gasoline. The toxicity of methanol, however, is commonly overstated, perhaps as a result of confusion with methyl-t-butyl ether (MTBE), a completely different chemical that, when used as an oxygenated gasoline additive, has caused significant groundwater contamination problems. In point of fact, methanol is present naturally in fresh fruit and vegetables, and so low doses of methanol have always been a normal part of the human diet. (It's also a major component of window washer fluid.) According to the FDA, a daily dose of 500 mg of methanol is acceptable for adult consumption.[7] This is good, because in addition to natural sources of methanol, many people today choose to consume diet soft drinks, which are sweetened with aspartame. Once inside the human body, aspartame is converted to methanol via the digestive process. The dose of methanol you could expect to get from inhalation during an auto refueling operation is about one-tenth as much as you receive from drinking a diet soda. If a large dose of methanol should be accidentally ingested, there is a widely available antidote: ethanol, which is preferentially taken up by the human body.

Neither methanol nor ethanol is cancer or mutation causing. In contrast, gasoline contains many carcinogens and mutagens, including benzene, toluene, xylene, ethyl benzene, and n-hexane.[8] As a result of fuel leaks and spills, incomplete combustion, and fumes from ordinary refueling operations, vast amounts of these gasoline carcinogens and mutagens are released into our environment every day, causing an increased incidence of cancer among the general public. The result is many deaths and billions of dollars in healthcare costs inflicted on the nation every year.

When burned in internal combustion engines, alcohol fuels do not produce any smoke, soot, or particulate pollution. According to the EPA, such pollution currently causes approximately forty thousand American deaths per year from lung cancer and other ailments. Converting to alcohol fuels could drastically reduce this toll.

Alcohols, especially methanol, also produce much less nitrogen oxide (NOx) pollution than gasoline, because they burn cooler. Since alcohols contain no sulfur, they produce no sulfur dioxide emissions at all. Thus, conversion to alcohol would also eliminate most of the vehicular contribution to acid rain.

Ozone smog is created when sunlight drives the reaction of nitrogen oxide and hydrocarbons in the atmosphere. As noted above, alcohol fuels produce less NOx than gasoline does. In addition, however, the reactivity of alcohol molecules (which might be released by incomplete combustion or evaporative emissions) with NOx in the atmosphere is less than a tenth as great as typical gasoline components.[9] Not only that, but because of their solubility in water, alcohol molecules are readily swept out of the atmosphere by rain.

Methanol can also be readily dehydrated to produce dimethyl ether—DME, chemical formula $(CH_3)_2O$.[10] Commonly used in aerosol spray cans, DME is an excellent diesel fuel with a cetane rating of 60. This compares quite favorably to about 45 to 50 for typical conventional diesel fuel. (The cetane rating is the measurement used to assess the quality of diesel fuel, in much the same way as the octane rating is used for gasoline.) Like the alcohols—and very much unlike conventional diesel fuel—DME produces no soot, particulate smoke, or sulfur dioxide, and very little of nitrogen oxides. Replacing conventional diesel fuel with DME would thus drastically cut air pollution from trucks, trains, ships, construction machinery, and portable stationary power generators. Such important additional air quality improvements would be an ancillary benefit of transitioning to a methanol-based fuel economy.

In the longer-term future, methanol might potentially be used to power fuel cell vehicles that have no pollution emissions of any kind. However, well before that, if used as gasoline and diesel replacements in ordinary cars and trucks, alcohol fuels and their derivatives could go a very long way toward cleaning up the air.

Ethanol is made from plant material and methanol can be. All fuel so produced acts in two ways as a counter to global warming. In the first instance, since plant material is derived from carbon dioxide drawn from the atmosphere, burning it produces no net CO_2 increase. Methanol made from natural gas that would otherwise be vented or flared, or from municipal waste that would otherwise be decomposed by microbes, is also global-warming neutral. In addition, however, the very act of growing plants acts as a powerful mechanism for active global cooling. This is so because the leaves of plants create an enormous amount of surface area for the transpiration and evaporation of water, in the process absorbing large amounts of heat from the environment. (That is why it feels cooler on a hot day to stand on the lawn rather than the pavement.) This heat is then incorporated into water vapor, which transports it high up into the stratosphere. When the vapor condenses, the heat is released, and most of it is lost to space.

The promotion of agriculture is thus the key to fighting global warming. This can most effectively be done through the alcohol economy, which will transfer trillions of dollars of business per year from the OPEC terror patrons to the world's farmers.

METHANOL AND AUTOMOBILE SAFETY

When discussing vehicle fuels, a key issue is safety. Fuels are, by definition, flammable, and thus, to a certain degree, intrinsically hazardous. However, alcohols in general are safer than gasoline, with methanol being the safest fuel of all.

Ethanol and methanol both have a much lower vapor pressure than gasoline, which means they produce much less combustible fumes. Not only that, but methanol fumes must be four times as concentrated as gasoline fumes in air for ignition to take place. As Nobel Prize–winner George Olah explains in his instructive book *The Methanol Economy*:

> If it does ignite, methanol burns about four times slower than gasoline and releases heat at only one-eighth the rate of gasoline fires. Because of the low radiant heat output, methanol fires are less likely to spread to surrounding ignitable materials. In tests conducted by the EPA and the Southwest Research Institute, two cars—one fueled by methanol and the other by gasoline—were allowed to leak fuel on the ground adjacent to an open flame. Whilst the gasoline ignited rapidly, resulting in a fire that consumed the entire vehicle within minutes, methanol took three times longer to ignite and the resulting fire damage affected only the rear part of the car. The EPA has estimated that switching fuels from gasoline to methanol would reduce the incidence of fuel-related fires by 90 percent, saving annually in the United States more than 700 lives, preventing some 4,000 serious injuries, and eliminating property losses extending to many millions of dollars.[11]

Unlike those fueled by gasoline, methanol and ethanol fires can be put out using water. The fact that they produce little smoke reduces the risk of smoke inhalation and makes fire fighting easier. Ethanol burns with a yellow fire that is clearly visible. Methanol has a blue flame that can be hard to see under some conditions, but this problem is easily dealt with by including an additive in the fuel. For example, by mixing 15 percent gasoline in the methanol (to make M85), a yellow flame is created that is quite visible even in bright sunlight.

It is for these reasons that methanol has been the fuel of choice for Indianapolis 500 race car drivers for more than three decades.

Alcohol flex-fuel vehicles are proven systems. They can be manufactured for the same price as gasoline cars, they pollute much less, and they significantly improve safety.

Moreover, they can function *today* in the real world, with existing infrastructure, while creating a vast market capable of driving global production of new cheaper fuels not controlled by our enemies.

It is its unique combination of these essential characteristics that makes flex-fuel technology the key practical weapon for victory in the energy war.

Unfortunately, the Bush administration has chosen a rather different approach. They call their plan "the hydrogen economy." However, as it shares virtually *none* of the essential requirements for a practical energy policy solution listed above, it would be more accurate to term it "the hydrogen hoax." As an energy battle plan for America, it guarantees our defeat. It needs to be debunked, as do other falsehoods currently blocking the way to implementation of a successful energy policy. In the next chapter, we will do just that.

CHAPTER 7

THE CHARLATANS STRIKE BACK

The Hydrogen and Pimentel Hoaxes

The invention and successful demonstration of alcohol flex-fuel vehicles has placed in our hands the necessary tool to switch the world off petroleum. Such a change would save Americans hundreds of billions of dollars per year, and humanity at large trillions per year. It would also provide the basis for a cleaner environment and open-ended, sustainable global economic growth that could lift billions of people out of poverty. Finally, it would destroy the financial power that is supporting the global spread of Islamofascism.

These are tremendous benefits. But where there are winners, there are losers, and in this case, those who stand to lose have many billions of dollars available to use in defense of their current privilege to loot the world. Thus it is hardly surprising to find various absurdities being vocally promoted in influential quarters for the purpose of deterring the adoption of an OPEC-smashing energy policy. The two most important of these are the hydrogen and Pimentel hoaxes. In this chapter, we will refute both of them.

THE HYDROGEN HOAX

> Yes my friends, I believe that water will one day be employed as fuel, that hydrogen and oxygen which constitute it, used singly or together, will furnish an inexhaustible source of heat and light, of an intensity of which coal is not capable . . . [when] the deposits of coal are exhausted we shall heat and warm ourselves with water. Water will be the coal of the future.
>
> —Jules Verne, *The Mysterious Island*, 1874

> I am inaugurating a program to marshal both government and private research with the goal of producing an unconventionally powered virtually pollution-free automobile within five years.
>
> —President Richard Nixon, State of the Union address, 1970

> President Bush's National Energy Policy laid out the vision for an economy run on hydrogen—the most common element in the universe. Hydrogen can fuel much more than cars and light trucks, our area of interest. It can also fuel ships, airplanes and trains. It can be used to generate electricity, for heating, and as a fuel for industrial processes.
>
> We envision a future economy in which hydrogen is America's clean energy choice—flexible, affordable, safe, domestically produced, used in all sectors of the economy, and in all regions of the country. . . .
>
> Imagine a world running on hydrogen later in this century: Environmental pollution will no longer be a concern. Every nation will have all the energy it needs available within its borders. Personal transportation will be cheaper to operate and easier to maintain. Economic, financial, and intellectual resources devoted today to acquiring adequate energy resources and to handling environmental issues will be turned to other productive tasks for the benefit of the people. Life will get better.
>
> —Secretary of Energy Spencer Abraham, November 12, 2002

> The sources of hydrogen are abundant. The more you have of something relative to demand for that, the cheaper it's going to be, the less

> expensive it'll be for the consumer. . . . Hydrogen power is also clean to use. Cars that will run on hydrogen fuel produce only water, not exhaust fumes. . . . One of the greatest results of using hydrogen power, of course, will be energy independence for this nation. . . . If we develop hydrogen power to its full potential, we can reduce our demand for oil by over 11 million barrels per day by the year 2040.
>
> —President George W. Bush, February 6, 2003

It certainly sounds great. We should use hydrogen for fuel. As then–energy secretary Spencer Abraham pointed out, it *is*, after all, the most common element in the universe. Since it is so plentiful, surely the president must be right when he promises it will be cheap. And when you use it, the waste product will be nothing but water! Imagine that, "water will be the coal of the future," just like Jules Verne said. "Environmental pollution will no longer be a concern." That's official, right from the top. Hydrogen will be abundant, cheap, and clean. Why settle for anything less?

Unfortunately, it's all pure bunk.

Hydrogen is only a source of energy if it can be taken in its pure form and reacted with another chemical, such as oxygen. But all the hydrogen on Earth, except that in hydrocarbons, already has been oxidized, so none of it is available as fuel. If you want to get unbound hydrogen, the closest place it can be found is on the surface of the sun. Mining this hydrogen supply would be quite a trick. (If you want to try, I suggest you go at night.) After the sun, the next closest source of free hydrogen would be the atmosphere of the planet Jupiter. Jupiter is surrounded by radiation belts so intense that they are deadly not only to humans, but even to electronics. It also has 318 times the mass of Earth, endowing the planet with a massive gravity field that would severely impair hydrogen export operations. These would also be complicated by the 2.5 year Jupiter-to-Earth flight transit time (during which all of the supercold liquid hydrogen launched from Jupiter would probably boil away), and the fact that upon reentry at Earth, the hydrogen shipping capsule would face heat loads about eight times higher than those dealt with by a space shuttle returning from orbit.

So if we put aside the spectacularly improbable prospect of fueling our planet with extraterrestrial hydrogen imports, the only way to get free hydrogen on Earth is to make it. Due to the fundamental laws of physics, in all cases, making hydrogen requires more energy than the hydrogen so produced would provide as fuel. Hydrogen, therefore, is *not* a source of energy. It simply is a carrier of energy. Furthermore, as we shall see, it is an extremely poor one.

The spokesmen for the hydrogen hoax like to sell their story by telling people that in the hydrogen economy, hydrogen will be manufactured from splitting water via electrolysis. It is certainly possible to make hydrogen this way, but it is very expensive—so much so that only 4 percent of all hydrogen currently produced in the United States is made in this manner. The rest is made through high-temperature breakdown (pyrolysis) of natural gas or steam reforming of hydrocarbons or coal.

Neither type of hydrogen is even remotely economical as fuel. The wholesale cost of commercial-grade liquid hydrogen (made the cheap way, from hydrocarbons) shipped in large quantities in the United States is about $6 per kilogram. For comparison, a kilogram of hydrogen contains about the same amount of energy as a gallon of gasoline. High-purity hydrogen made from electrolysis for scientific applications costs considerably more. Dispensed in compressed gas cylinders to retail customers, the current price of commercial-grade hydrogen is about $100 per kilogram.

Thus, unlike methanol, which can be made for less than gasoline, or ethanol, which is roughly competitive, hydrogen costs much more than conventional fuels. This means that even if hydrogen cars were available—and hydrogen stations existed to fuel them—no one with the power to choose otherwise would ever buy such vehicles. This fact alone makes the hydrogen economy a nonstarter in a free society.

If, however, you are among those willing to sacrifice freedom and economic rationality for the sake of the environment, and therefore prefer hydrogen for its advertised benefit of reduced carbon dioxide emissions, think again. Because hydrogen is actually made by reforming hydrocarbons, its use as fuel would not reduce greenhouse gas emissions at all. In fact, it would greatly increase them.

To see this, let us consider an example. Let's say you wanted to produce hydrogen. You choose to do it via steam reformation of natural gas, the most common technique used commercially today. The reaction is endothermic, and will need an outside source of energy to provide the 59 kcal/mole needed to drive it forward. As discussed in chapter 5, this can be obtained by burning some methane, which releases 205 kcal/mole.

Assuming an optimistic 72 percent efficiency in using the combustion energy to drive the steam reformation, this would allow us to reform 2.5 moles of methane for every 1 that we burn (or 5 for every 2). So if we take 5 units of the methane steam reformation reaction and add it to 2 units of methane combustion, the net reaction becomes:

$$7CH_4 + 4O_2 + 6H_2O \rightarrow 7CO_2 + 20H_2 \quad (1)$$

So, as far as fuel is concerned, what we have managed to do is trade 7 moles of methane for 20 moles of hydrogen. Seven moles of carbon dioxide have also been produced, exactly as many as would have been produced had we simply used the methane itself as fuel. The 7 moles of methane that we used up, however, would have been worth 1,435 kcal of energy if used directly, while the 20 moles of hydrogen we have got in exchange for all our trouble are only worth 1,320 kcal. So for the same amount of carbon dioxide released, less useful energy has been produced.

The situation is much worse than this, however, because before the hydrogen can be transported anywhere, it needs to be either compressed or liquefied. To liquefy it, it must be refrigerated down to a temperature of –253°C (20 degrees above absolute zero). At these temperatures, fundamental laws of thermodynamics make refrigerators extremely inefficient. As a result, about 40 percent of the energy in the hydrogen must be spent to liquefy it. This reduces the actual net energy content of our product fuel to 792 kcal. In addition, because it is a cryogenic liquid, still more energy could be expected to be lost as the hydrogen boils away as it is warmed by heat leaking in from the outside environment during transport and storage.

As an alternative, one could use high-pressure pumps to compress the hydrogen as gas instead of liquefying it for transport. This would require wasting only about 20 percent of the energy in the hydrogen to accomplish. However, safety-approved steel compressed gas tanks capable of storing hydrogen at 5,000 pounds per square inch (psi) weigh approximately sixty-five times as much as the hydrogen they can contain. So to transport 200 kg of compressed hydrogen, equal in energy content to just 200 gallons of gasoline, would require a truck capable of hauling a 13 tonne load. Think about that: An entire large truckload delivery would be needed simply to transport enough hydrogen to allow *ten* people to fill up their cars with the energy equivalent of 20 gallons of gasoline each.

Instead of steel tanks, one could propose using (very expensive) lightweight carbon fiber overwrapped tanks, which weigh only about ten times as much as the hydrogen they contain. This would improve the transport weight ratio by a factor of six. Thus, instead of a 13-tonne truck, a mere 2-tonne truckload would be required to supply enough hydrogen to allow a service station to provide fuel for ten customers. This is still hopeless economically, and could probably not be allowed in any case, since carbon fiber tanks have low crash resistance, making such compressed hydrogen transport trucks deadly bombs on the highway.

At enormous expense, a system of pipelines could, in principle, be built for transporting gaseous hydrogen. However, because hydrogen is so diffuse, with less than one-third the energy content per unit volume as natural gas, these pipes would have be very big, and large amounts of energy would be required to move the gas along the line. Another problem with this scheme is that the small hydrogen molecules are brilliant escape artists. Hydrogen not only penetrates readily through the most minutely flawed seal, it can actually diffuse right through solid steel itself. The vast surface area offered by a system of hydrogen pipeline would thus afford ample opportunity for much of the hydrogen to leak away during transport.

As hydrogen diffuses into metals, it also embrittles them, causing deterioration of pipelines, valves, fittings, and storage tanks used throughout the entire distribution system. These would all have to be monitored, inspected, tested, and replaced on a continual basis; otherwise the distribution system would become a source of continuous catastrophes.

As a consequence of these technical difficulties, the implementation of an economically viable method of retail hydrogen distribution from large-scale central production factories is essentially impossible. Because of this, an alternative concept has been proposed wherein methane or methanol fuel would be transported by pipeline or truck, and then be steam reformed into hydrogen at the filling station itself. This would eliminate most of the cost of hydrogen transport, but would increase the cost of the hydrogen itself, since small-scale reformers are less efficient, both economically and energetically, than large-scale industrial units. Also, it is questionable how many service stations would want to pay to buy, operate, and maintain their own steam-reforming facilities. Each station would also need to operate its own 5,000 psi explosion-proof high-pressure hydrogen pump or a cryogenic refrigeration plant, both of which are very unappealing prospects. Such a scheme of distributed production stations also would eliminate any hope of implementing the hydrogen economy's advertised plan to eventually sequester underground the carbon dioxide produced as a by-product of its hydrogen-manufacturing operations. It also would be fundamentally ridiculous, since either the methane or the methanol used as feedstock at the station to make the hydrogen would be a better automobile fuel, containing more energy, in less volume, at less cost, than the hydrogen it yields.

The idea of producing hydrogen via water electrolysis locally at filling stations is equally preposterous. To see this, consider the following: A kilogram of hydrogen has the same energy content as a gallon of gasoline, so the owner of a filling station could only expect to obtain the same net income from a kilogram of hydrogen as from a gallon of gas. A reasonable figure for this might be $0.20 per kilogram. To obtain a modest net income of $200 per day from hydrogen sales would therefore require selling 1,000 kilograms per day. Since hydrogen requires about 163,000 kJ/kg to produce via electrolysis (assuming an 85 percent efficient electrolyzer), this means that 163,000,000 kJ = 45,278 kW-hours per day would be required by the station. At current grid power costs of $0.06 per kW-hour, this would run the station an electric bill of $2,717 per *day*. If the electrolysis unit ran round the clock, it would need to be supplied with 1,900 kW of electricity (about a thousand times the power draw of a typical house). This power would need to be supplied by the utility over special heavy-duty lines, and then transformed and rectified into direct current on site for use in the electrolyzer. Electrolyzers use high-amp/low-voltage power. In this case, at least *several hundred thousand amps* would be required. The 1,900 kW electrolyzer would not be cheap either. At current prices such a unit would run the station owner more than $10 million, which, mortgaged over thirty years, would cost him about $100,000 per month, assuming it lasted that long. (Why anyone would want to do this is beyond me, since the same $10 million put into 5 percent bonds would return $500,000 per year, or ten times the $200 per day hydrogen sales income under discussion, with no work and no risk.) Then, of course, he would still need to buy and operate either a 5,000 psi explosion-proof compressor pump or a cryogenic refrigerator, and build and accept liability for high-pressure or cryogenic hydrogen storage facilities on his properties. Having paid for all that, there would then be the little matter of insurance.

Such a deal. For just $6,000 per day, plus insurance costs, you could make $200, provided you can find fifty customers every day willing to pay triple the going price for automobile fuel. I don't know about you, but if I were running a 7-Eleven, I'd find something else to sell.

The writer Lewis Carroll once said that he could believe ten impossible things before breakfast. Such an attitude is necessary in discussing the hydrogen economy, since no part of it is possible. Therefore, putting aside the intractable issues of fundamental physics, hydrogen production costs, and distribution showstoppers that we have raised thus far, let us proceed to discuss the problems associated with the hydrogen cars themselves.

In order for hydrogen to be used as fuel in a car, it has to be stored in the car. As at the station, this could be done either in the form of super-cold liquid hydrogen or as highly compressed gas. In either case, we come up against serious problems caused by the low density of hydrogen. For example, if liquid hydrogen is the form employed, then storing 20 kg onboard (equivalent in energy content to 20 gallons of gasoline) would require an insulated cryogenic fuel tank of some 280 liters (70 gallons) volume. This cryogenic hydrogen would always be boiling away, which would create concerns for those who leave their cars parked for any length of time, and which would also turn the atmospheres in underground or otherwise enclosed parking garages into fuel-air explosives. Public parking garages containing such cars could be expected to explode regularly, since hydrogen is flammable over concentrations in air ranging from 4 percent to 75 percent, and the minimum energy required for its ignition (0.005 mcal) is about one-twentieth that required for gasoline or natural gas.[1]

If, instead, 5,000 psi compressed hydrogen were employed, the tank would need to be 650 liters (162 gallons), or eight times the size of a gasoline tank containing equal energy. Because it would have to hold high pressure, this huge tank could not be shaped in an irregular form to fit into the vehicle's empty space in some convenient way. Instead it would have to be a simple shape like a sphere or a domed cylinder, which would make its spatial demands much more difficult to accommodate and significantly reduce the usable vehicle space within a car of a given size. If made of (usually) crash-safe steel, the hydrogen tank would weigh 1,300 kg (2,860 lb)—about as much as an entire small car! Lugging this extra weight around would drastically increase the fuel consumption of the vehicle, possibly by as much as a factor of two. If instead a lightweight carbon fiber overwrapped tank were employed to avoid this penalty, the car would become a deadly explosive firebomb in the event of a crash.

Hydrogen gas can be used as a fuel in internal combustion engines, but there is no advantage in doing so. In fact, in contrast with methanol, which enables a 15 percent increase in internal combustion engine efficiency due to higher compression ratios, hydrogen reduces the efficiency of such engines by 20 percent compared to what they can achieve using gasoline. For this reason, nearly all discussion of hydrogen vehicles has centered around power systems driven by fuel cells.

Fuel cells are electrochemical systems that generate electricity directly through the combination of hydrogen and oxygen in solution. Essentially electrolyzers operating in reverse, they are attractive because they have no moving parts (other than small water pumps), are quiet, and, under conditions where the quality of their hydrogen and oxygen feed can be perfectly controlled, are quite efficient and reliable. These features have provided sufficient advantages to make fuel cells the number-one technology of choice for certain specialty applications, such as the power system for NASA's *Apollo* capsules and the space shuttle.

However, despite their successful use for four decades in the space program, and many billions of dollars of research and development funds expended over the years for their improvement and refinement, fuel cells have thus far found little use in broader terrestrial commercial applications. The reasons for this are threefold. In the first place, for applications here on Earth, a practical power system must last years, not just the few weeks required to support a manned space flight. In the second place, on Earth, the oxygen supply for the fuel cell must come from the atmosphere, which contains not only nitrogen (which decreases the fuel cell efficiency compared to a pure oxygen source), but also carbon dioxide, carbon monoxide, and many other pollutants, which even in trace form can contaminate the catalysts used in the fuel cells, causing their permanent degradation and ultimately rendering the system inoperable. Finally, and decisively, fuel cells are very expensive. For NASA, which spends a billion dollars on every shuttle launch, it makes no difference if its 10 kW fuel cell system costs $100,000, $1 million, or $10 million. For a member of the public, however, such costs matter a great deal.

There are many kinds of fuel cells, including alkaline, phosphoric acid, and molten carbonate systems, but for purposes of motor vehicle use the only kind that is suitable and which is being pursued for development is the proton exchange membrane fuel cell (PEMFC). These, for example, are the kind used by all vehicle fuel cell engines manufactured by the Ballard Power Company of Vancouver, British Columbia, which for the past decade has been responsible for about 80 percent of all fuel cell engines produced worldwide.

PEMFCs use a platinum catalyst, which is very expensive, and despite billions of dollars of research and development efforts to reduce the amount required, it has proven impossible to cut the cost of such systems below about $7,000 per kW.[2] This is very unfortunate, because an electric car with a 100 horsepower motor needs about 75 kW of electricity to make it go. At this price, the cost *for just the fuel cell stack* powering the car would be about half a million dollars. Actual costs for *complete* Ballard fuel cell engine systems have been well over a million dollars each.[3] Then there's still the rest of the car to pay for, although with the propulsion system costing this much, that's pretty much in the noise.

That, however, is not even the worst of it. Operating under road conditions in the real atmosphere, which contains such powerful catalyst poisons as sulfur dioxide, nitrogen dioxide, hydrogen sulfide, carbon monoxide, and ammonia, which can permanently incapacitate a PEMFC, the operating lifetime of fuel cell stacks has been shown to be less than 20 percent those of conventional diesel engines. As the trenchant independent industry analyst F. David Doty pointedly put it:

> We're still waiting to see a fuel-cell vehicle driven from Miami to Maine via the Smoky Mountains in the winter—even one time, with a few stops and restarts in Maine. Then, we need to see one hold up to a 40-minute daily commute for more than two years (preferably at least 15 years) with minimal maintenance, and come through a highway accident with less than $200,000 in damages. . . .

> When lifetime and maintenance are considered, one can argue that vehicle-qualified PEMFCs are currently 400 times more expensive than diesel engines.[4]

It is true that the cost of PEMFCs might conceivably be reduced over time as a result of technology improvements (although no progress in manufacturing cost reduction has been achieved over the past decade—despite several billions in research investment). However, if somehow the vehicles ever went into mass production, increased demand would drive up the price of the platinum they contain, and thus the overall system cost, right through the roof.

The outrageously high costs of both fuel cell cars and hydrogen fuel, combined with the nonexistence of a hydrogen distribution and sales infrastructure and the danger to life and limb involved in driving around a vehicle containing a crash-detonatable hydrogen gas bomb, make the possibility of mass consumer purchases of hydrogen fuel cell vehicles a nonstarter. But let's say some benevolent government bureaucrat with a great deal of your money to spend decided to take $700 billion out of the treasury to buy, at $1 million each, 700,000 PEMFC powered vehicles (this would represent about four-tenths of 1 percent of the US total fleet), and then spring for another $300 billion to establish a hydrogen distribution infrastructure. Admittedly, it's a lot of money, and the resulting highway incinerations would be unfortunate, but wouldn't we at least get some environmental benefit for our trillion bucks?

No, we would get no benefit at all. As discussed above, hydrogen is actually produced commercially using fossil fuel energy, much of which is lost in the process, meaning that *more* fossil fuels need to be burned, and thus *more* carbon dioxide produced, to provide a vehicle with a given amount of energy using hydrogen than if the vehicle were allowed to burn fossil fuels directly. Even if we ignore costs completely and generate hydrogen for vehicle fuel using water electrolysis, that would also increase pollution, since most electricity is actually generated by burning coal and natural gas. Even if the electricity in question came from nuclear, hydro, wind, or solar power, wasting it on hydrogen generation would also increase overall societal carbon dioxide emissions relative to the simpler alternative of putting the power into the grid.

Furthermore, despite all their cost and hype, the fuel cell vehicles themselves offer no increase in efficiency relative to more conventional systems. While the theoretical efficiency of a hydrogen/oxygen fuel cell approaches 85 percent, the actual efficiency of real PEMFC stacks using hydrogen and air near maximum output (where they must operate because fuel cell capacity is so expensive) is about 38 percent.[5] If we then factor in an estimated efficiency for the power electronics of 92 percent and a real-world motor efficiency of 85 percent, we obtain an estimate of about 30 percent efficiency for a fuel cell vehicle. Ordinary internal combustion engine cars can already match this, with systems offering up to 38 percent efficiency well in sight. Modern conventional diesel engines operate *today* at about 42 percent efficiency. With variable valve timing, they should be able to attain 58 percent efficiency.[6] That's nearly *twice* the efficiency offered by a fuel cell vehicle, at 1/400th the cost.

In short, hydrogen vehicles offer no practical or societal advantages whatsoever.

Despite this, the US Department of Energy has continued to hand out billions of dollars of taxpayers' money to major auto companies and their fuel cell development partners to produce hydrogen-powered auto show display vehicles, and issue repeated predictions claiming that tens of thousands of these cars will soon be appearing on America's highways. In fact, the DOE's repeated past projections of the growth of hydrogen vehicles have all been at least two orders of magnitude higher than reality. As a result, stocks in all the major fuel cell companies, pumped high by such hype at the expense of naive investors, are currently selling at less than one-tenth their peak past posthype values.

To summarize, the hydrogen economy is a nonstarter. The infrastructure to refuel hydrogen vehicles does not exist, so no hydrogen cars will be bought by anyone, since there is nowhere to refuel. No hydrogen-refueling infrastructure can develop because there are no hydrogen cars, and so no transition is possible. Even if a distribution system did exist, however, no one would buy a hydrogen car anyway, because both the cars and their fuel cost far too much, and they can't store enough hydrogen on board to drive anywhere. There would be no societal advantage to using hydrogen in any case, because the inefficiency associated with its production means that much more energy, and thus much more pollution, is needed to produce it than any other fuel. Its inefficiency of production and use also means that switching to hydrogen would increase rather than decrease America's dependence upon foreign oil. Furthermore, since hydrogen cannot be economically transported long distances, a hydrogen economy would offer none of the alcohol economy's prospects for mobilizing the resources of the third world to meet advanced-sector energy needs through mutually beneficial trade.

The "hydrogen economy" is thus a hoax. If we are to win the energy war, we need to shun such hallucinations and adopt a practical plan that can really work. That is the alcohol economy.

THE PIMENTEL HOAX

Ethanol is an excellent automotive fuel. In the pure form, it is 113 octane, outclassing even methanol's 105. It is nearly as safe and nonpolluting as methanol, and, uniquely among vehicle fuels, it is not only nontoxic, it is actually edible. The energy content of ethanol is only two-thirds as much as gasoline per gallon, so fuel tanks would need to be 50 percent larger to get the same range on a fill-up. But since, unlike hydrogen (but like methanol), ethanol is a liquid at room temperature, it can be stored in ordinary unpressurized fuel tanks that can be shaped to fit any available space in the vehicle. As a result, accommodating the modest extra fuel volume required is not a problem. Ethanol mixes well with gasoline and has been proven to operate well in millions of existing flex-fuel cars that cost little or nothing more than their gasoline-only counterparts. Ethanol is made from the sugar and starch content of agricultural products, a sector where the United States and its allies are quite strong. By buying ethanol we can redirect funds, which would otherwise go to the Saudis to promote terrorism, to our own farm sector or that of our hemispheric neighbors.

The only downside to ethanol is its cost. Currently, ethanol made from US corn costs about $1.50 per gallon to produce. This is the energy equivalent to gasoline at $2.25 per gallon, so when gas retail prices are $3.00 per gallon (about $2.50 per gallon before taxes) or higher, corn-derived ethanol is quite competitive. However, oil prices fluctuate considerably in accord with OPEC's whims and tactics, so it has been frequently the case that gasoline has been cheaper.

In order to cope with this situation and allow a US ethanol fuel industry to take root and grow, the Carter administration in 1980 instituted a program providing a 40-cents-per-gallon subsidy for agricultural ethanol production. Strongly supported by the US farm lobby, this subsidy has remained in place, being raised slightly over the years to its current level of 51 cents per gallon. So long as oil prices remained low during the 1980s and 1990s, this program induced only a modest rate of growth of ethanol fuel production. However, following OPEC's action in jacking up oil prices since 1999—and especially in 2001—subsidized corn ethanol became compellingly profitable and the industry took off. This is shown dramatically in figure 7.1, where the sharp increase in the industry growth rate since 2001 is clear.

It must be said that despite this strong growth, US ethanol production is not yet large enough to make a major impact on energy security. The 4.9 billion gallons produced in 2006, equivalent in energy to about 3.3 billion gallons of gasoline, represented only about 2.5 percent of the total US vehicle consumption of 131 billion gallons of gasoline for the year. Still, it is a start that promises substantially more, especially once flex-fuel vehicles are put on the road in large numbers. Furthermore, and very important, the growing ethanol industry has finally created, in the form of the powerful American

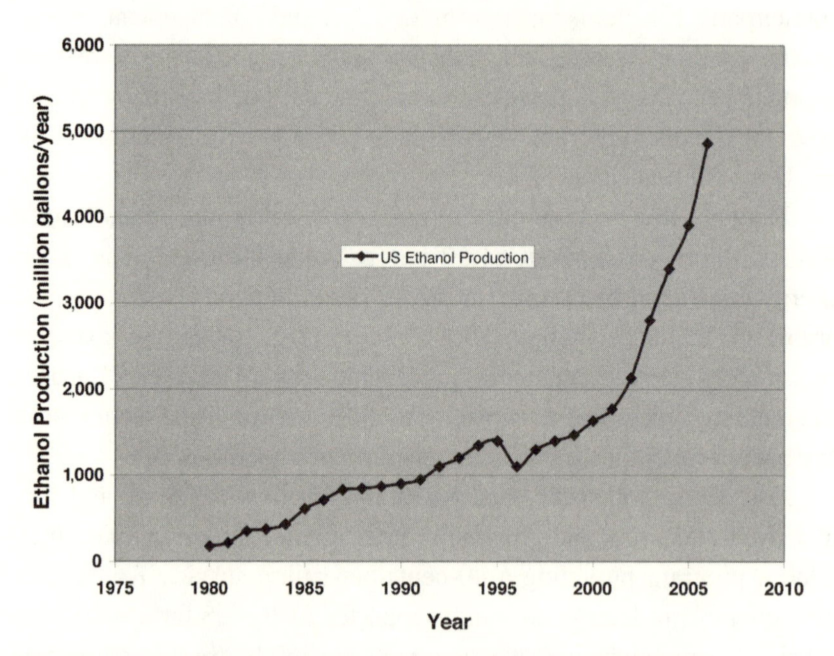

Figure 7.1. Corn ethanol production in the United States, 1980–2006.
***Source*: Energy Information Administration, http://www.eia.doe.gov.**

farm lobby, a political champion for alternative fuels that might just have the strength to go toe-to-toe against the oil cartel.

Not everyone is happy with this development, of course, and the reasons are plain to see. The 4.9 billion gallons of US ethanol produced in 2006 took $10 billion away from the oil cartel. Thus it is hardly surprising to find the ethanol program regularly denounced by journalistic hired guns and other business analysts associated with oil industry–funded think tanks, as well as by ideological libertarians whose sensibilities it offends.

One of those who have been especially vocal in this regard is American Enterprise Institute (AEI) resident scholar Dr. Kevin Hassett. According to Hassett, a former Columbia University professor who served as John McCain's chief economic adviser during his 2000 presidential campaign, the ethanol program is a "scam" that violates all sound principles of free-enterprise economics as laid down by Adam Smith.

Stated thus far, Hassett's arguments are simply standard-issue laissez-faire fundamentalism, although slightly disingenuous since his antiethanol subsidy polemics omit mention of the much larger government subsidies provided to the oil industry (more than $180 billion over the past decade). However, Hassett—who is, incidentally, the coauthor of the 1999 business classic *Dow 36,000: The New Strategy for Profiting from the Coming Rise in the Stock Market*[7]—takes another step. Not only are ethanol subsidies an illegitimate use of tax dollars, he says, but the program is worthless as a step toward energy independence because it takes more fossil fuel to produce ethanol than the process yields.

Hassett is a capable writer and his polemics sound powerful, but are they correct? (*Dow 36,000* wasn't.) Certainly the ethanol subsidy program violates the free market, but does it reduce or increase our dependence upon foreign oil? At the current subsidy rate of $0.51 per gallon, the nation (which now spends about $2.4 billion per year on ethanol subsidies) could replace the entirety of the gasoline we derive from Mideastern oil (20 percent of our total) at a cost of $20 billion per year. That is a very small fraction of the hundreds of billions of defense expenditures the nation is forced to pay defending its access to Arab oil, to say nothing of the additional costs incurred by those terrorist states and organizations funded by the roughly $100 billion in foreign oil purchases the said ethanol subsidy would eliminate. So while libertarians may wince, if the ethanol program actually does reduce foreign oil dependency, then it may well be worthwhile.

Hassett, however, says it does not. In his widely circulated AEI paper on the matter, he provides as a source for this claim "a recent careful study by Cornell University's David Pimentel and the University of California at Berkeley's Tad Patzek."[8] Now, former Shell executive Patzek is an oil industry man with, accordingly, no claim to objectivity or credibility on the ethanol issue, but Pimentel is a noted insect ecologist without any vested business interest to defend. Surely a study written primarily by the distinguished Professor Pimentel must be taken as authoritative.

Well, actually, no. The "recent careful" Pimentel-Patzek study was published in 2001,[9] and was virtually immediately (in 2002) shown to be grossly erroneous in a review performed by Dr. Bruce Dale, a professor of chemical engineering at Michigan State University.[10] Dale's critique was harsh, clear, and forceful. Among the errors Dale identified in Pimentel's study:

- Pimentel's corn yields date from 1992 (and are thus underestimated).
- Pimentel's figures for energy required to produce ethanol and the ethanol yield date from 1979, and his figures for energy to produce fertilizer are 1990 world values per the UN Food and Agricultural Organization (FAO)—not recent US values (and thus grossly overestimated).
- Pimentel assumes all corn is irrigated (only 16 percent is, and virtually no irrigated corn is converted to ethanol). This last point is a source of very large error, since Pimentel assigned huge energy costs for ethanol corn crop irrigation.
- Pimentel fails to assign any energy credit for the high protein animal feed produced as a by-product of ethanol production. (Most of the protein value of the corn crop used for ethanol is preserved, and used as animal feed to produce meat. If the ethanol were not being produced, most of the energy required to grow this feed would have to be expended anyway.)

In conclusion, the Dale study showed that not only was the *energy* balance for producing ethanol significantly positive—not negative, as Pimentel and Patzek had claimed—but the much more relevant (for energy independence) metric of balance of the amount of *liquid fuel* produced versus that expended in its production was enormously favorable. In fact, said Dale, at least six gallons of ethanol are produced for every gallon of gasoline or diesel fuel expended in the process. He later revised this to "more than twenty gallons."

Dale's thorough refutation of the Pimentel report is not an obscure document. It was presented in testimony before the US Congress on July 31, 2002, and has been widely cited in professional literature. It thus may be considered odd that neither Hassett nor a host of other "pro-free enterprise" antiethanol journalists ever take notice of it in their articles, which continue to recycle the Pimentel line.

Perhaps one might view the Dale-Pimentel dispute as simply a matter of battling professors, with no way for an outsider to decide the merit of the case. Arguably, this assessment might leave someone like Hassett free to choose his favorite contender, selecting his fellow Ivy Leaguer from Cornell over the critic from plebian Michigan State.

Actually, however, it is easy to see who is right. Ethanol currently sells for about $1.50 per gallon wholesale before the $0.51 per gallon subsidy, leaving the farmer only about $2 per gallon of gross income for every gallon of ethanol sold. Gasoline and diesel fuel both sell for well over $2 per gallon. Thus, even if we ignore *all* other production costs, it is *mathematically impossible* for the production of a gallon of ethanol to require more than a gallon of liquid fuel. If we include these other costs—such as land, labor, equipment, repairs, taxes, and so forth—that comprise the large majority of a farmer's budget, it is clear that the fuel required to produce ethanol must be very small compared to the product yield.

Thus it is readily apparent to any numerate person who takes the trouble to consider the problem that the Dale report was substantially correct, and the Pimentel study not only erroneous but ludicrous.

For the benefit of those who can't count for themselves, the final nail in the coffin of Pimentel's credibility was hammered home in a refereed paper published by Alex Farrell and his colleagues at the Berkeley Energy and Resources Group in *Science* magazine in January 2006.[11] Their key result is shown in figure 7.2, taken from the paper, reproduced here. The authors went over the calculations of all the prior literature, including Pimentel and Patzek's, and considered the critical question of how much petroleum is expended by making a given amount of liquid fuel energy in the form of ethanol compared to that used to make the same fuel energy's worth of gasoline. Even using

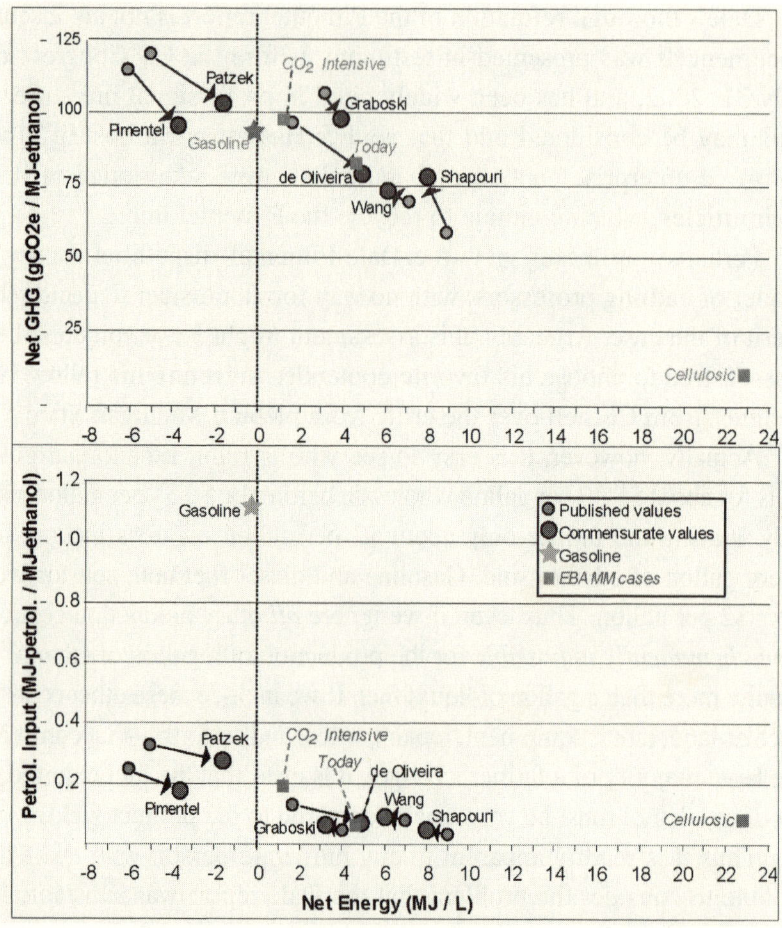

Figure 7.2. Graphs taken from the important paper by Farrell et al. Fig A (top) shows that using ethanol reduces greenhouse gas (GHG) emissions by about 25 percent. Fig. B (bottom) shows that using ethanol reduces petroleum requirements by a factor of 10. *Source*: A. Farrell, R. Plevin, B. Turner, A. Jones, M. O'Hare, and D. Kammen, "Ethanol Can Contribute to Energy and Environmental Goals," *Science* 311 (2006): 506–508. Reprinted with permission from AAAS.

Pimentel's assumptions, they found that the petroleum cost of ethanol fuel was about one-fifth that of gasoline. Based upon the assumptions of nearly all other analysts, the petroleum cost of ethanol was less than one-tenth of gasoline. In other words, for a given amount of petroleum used, more than ten times as much ethanol can be produced as gasoline. Therefore, by switching to ethanol, we can reduce the amount of petroleum needed to make our fuel by as much as 90 percent! Pimentel's claims could not be further from the truth.

Who Is David Pimentel?

But how could such a distinguished professor from such a distinguished university be so wrong? This is a very interesting question, the answer to which must surely have horrified Hassett and any other Pimentel-quoting free enterpriser who bothered to look into the matter.

Pimentel is not just an opponent of ethanol production. He is also an opponent of beef production. He is an opponent of the use of pesticides and of modern agriculture in general, which he has attacked with an endless stream of defective papers since 1974.[12] He denounces industry as well, in one paper making the wild claim that "we have calculated that an estimated 40 percent of world deaths can be attributed to various environmental factors, especially organic and chemical pollutants."[13] He is also highly critical of housecats and pet dogs, which he deplores as "alien species" introduced into North America. He's against human immigrants, too—both legal and illegal. And then there are babies. Pimentel believes there should be fewer of them. Much fewer.

David Pimentel is a radical Malthusian. According to Pimentel, the earth has a maximum "carrying capacity" of 2 billion people, and the United States of 100 million people, and the populations of both must be reduced accordingly. This, according to Pimentel, can best be accomplished by "democratically determined population control practices."[14] Using such methods, he says, the average birthrate can be reduced to 1.5 children per couple. The average standard of living in the United States will also have to be cut in half. "None of these solutions, unfortunately, will be painless," says Pimentel.[15] No doubt.

In 2004 Pimentel ran for the board of the Sierra Club on a joint slate with former Colorado governor Richard Lamm, himself noteworthy for his advocacy of early life termination of the elderly.[16] The platform of the Pimentel-Lamm slate was to save the American environment by putting a halt to *all* immigration, and their campaign promptly received the enthusiastic support of a broad array of extremist organizations, including some with openly racist or neo-Nazi affiliations. This situation so alarmed the leadership of the Sierra Club that Executive Director Carl Pope put out a general alert to the membership to stop the takeover of the organization by "racist hate groups." Some of Pimentel's more ardent fans responded by branding his Sierra Club opponents with anti-Semitic epithets.[17]

Of course, the fact that Pimentel's supporters include neo-Nazis doesn't necessarily make him one. However, as documented by journalist Betsy Hartmann in a *New Scientist* interview titled "The Greening of Hate,"[18] the distinguished professor from Cornell seems to place himself repeatedly on the boards of various obscure journals and organizations in company with some very questionable figures. For example, Pimentel is a member of the board of directors of the extreme Malthusian anti-immigrant Carrying Capacity Network (CCN). Self-proclaimed "white separatist" Virginia Abernethy is the CCN's chairman of the board.[19] With respect to the energy issue, the CCN's viewpoint is simple: Our problem is not that we don't have enough fuel, but that we have too many people. Get rid of the extra people, and the problem will be solved.[20]

Environmentalists come in two varieties: prohuman and antihuman. Prohuman environmentalists seek practical solutions for real problems in order to enhance the environment for the benefit of humanity. Antihuman environmentalists seek to make use of instances of inadvertent human damage to nature as an ideological weapon on behalf of the age-old reactionary thesis that humans are nothing but vermin whose aspirations need to be contained and suppressed by tyrannical overlords to preserve a divinely ordained static natural (and social) order. This is the fundamental premise of Malthusian ideology, whose twentieth-century descendants notably include not only antihuman environmentalism but also Nazism. The same bestialist view of human nature is also the ultimate root of Islamofascism, so, as we will discuss in greater detail in the final chapter, it's not surprising to find them all in the same camp.

The Sierra Club includes both types of environmentalists, as well as many people who join it simply to participate in its outdoor activities. By naming humans as such, rather than specific human activities, as the enemies of nature, the Pimentel-Lamm campaign was an attempt to purge the Sierra Club of its humanistic currents and turn the nation's flagship environmental organization into a fascist juggernaut. Fortunately, the antihuman ticket lost.

The Free Marketeers' Responsibility

Pimentel opposes economic growth; Hassett and his free market friends would like to see more of it. Pimentel is the darling of extremist anti-immigrant agitators; Hassett has published articles advocating a moderate immigration policy. Pimentel advocates "democratic" but nevertheless state-ordained population control; Hassett, as a classical liberal follower of F. A. Hayek,[21] could not be suspected of supporting such a program, and I have little doubt that he must regard the rest of Pimentel's social ideas with an appropriate degree of revulsion. Without question, Hassett chooses to trot out Pimentel not because he is a Pimentelist, but because he sees an advantage in recycling a discredited study to make a political point.

Yet, aside from the issues of academic honesty and rigor involved in engaging in such a procedure, there are serious questions that Hassett and the other Pimentel-parroting pro-OPEC "libertarians" need to answer. For example, given Pimentel's public prominence resulting from his Sierra Club campaign, is it really possible that these people do not know they are citing a Malthusian zealot? And given the vital importance of the energy independence issue for the national security and well-being of the United States, is it really appropriate to deal with it using such fakery?

As the Bible says, a rotten tree cannot bear good fruit. Pimentel's analysis is pure bunk. Having cleaned the stable, it's time for us to move on.

CHAPTER 8

THE NEW ALCOHOL ECONOMY AND WORLD DEVELOPMENT

One of the last places a camera crew and I went to complete a documentary in South Africa was a squatters camp where people can't even imagine having sanitation or fresh water or easily available electricity. . . . They can't conceive of life with less than 106 people for each concrete outhouse toilet. They can't because many were born in these camps, grew up in them, made love in them, and will die in them.

Millions of South Africans live in these makeshift slums euphemistically called "informal settlements," either in tin shacks, most no bigger than my Colorado kitchen, or in tiny "bungalows" of rough wooden slats through which the wind whistles almost without obstruction. We met families of up to twelve people, sleeping side by side on two adjacent single beds that occupy about 90 percent of the floor space. They cook over propane flames, and have only candles each night to relieve the dank darkness.

I asked one woman how long she and her family had lived this way. "For years," she said. I asked, "How will you ever get someplace better?" She answered, "By getting work." I asked, "Have you looked?" She said, "For years."

—Greg Dobbs, *Rocky Mountain News*, September 2, 2006

The greatest problem in the world today is poverty. Despite all the advances we have made, billions of people around the globe still live in conditions that are morally unacceptable. Every year, tens of millions die of starvation or disease, while myriads more are blinded or crippled. Whole populations live in squalor and ignorance, denied all but the tiniest crumbs of the material and intellectual fruits of twenty-first-century civilization. Nearly 2 billion people remain illiterate. In the nonindustrialized world as a whole, combined unemployment and underemployment averages 30 percent. While the worldwide per capita gross domestic product (GDP) is about $8,800, nearly half of the global population is forced to live on incomes under $2 per day.[1] In the fifty poorest countries, life expectancy is under forty years, and infant mortality exceeds 10 percent.[2]

One could go on at length, citing figures documenting the lack of education, sanitation, clean drinking water, housing, heating, electricity, medicine, nutrition, and opportunity, alongside numerous other scarcities that continue to torment, degrade, and destroy the lives of the majority of humanity. But I don't think that is necessary. While most of us try to avoid looking at the unpleasant picture, we all know what is going on. It is the grim reality, the dirty underside of our gilded age.

Of course, being people of conscience, we, the voters of the advanced nations of the West, have not ignored this issue. By significant majorities we have, for decades, elected governments committed to alleviating the problem of third world poverty. Many of us also choose to contribute directly to projects initiated by churches or independent nonprofit development groups. As a result, some $60 billion in Western aid is currently being sent to the underdeveloped sector every year. Over the past fifty years, some $2.3 trillion (in today's money) of development aid has been sent, including about a third from the United States.

Unfortunately, it hasn't worked.[3] While there have been a few large-scale success stories due to actions of the major donor organizations, including the elimination of polio in Latin America and measles in southern Africa, and many good small-scale projects implemented by the private aid groups, the general situation has not changed. Half the world still lives in misery.

Two schools of thought have appeared to explain the failure. One, the liberal group typified by former British prime minister Tony Blair, various World Bank officials, and Columbia University professor and UN Millennium Project director Jeffrey Sachs,[4] argue that the amount of aid simply hasn't been enough. The other, more conservative group, exemplified by New York University professor and Center for Global Development fellow William Easterly, counter that most of the aid money has been wasted, and that such gross waste is an intrinsic feature of the foreign aid process.

Unfortunately, both groups appear to be right. The liberals are right because, as substantial as it might appear, $60 billion per year is really a meager budget if the problem to be addressed is third world poverty. There are 3 billion people on this planet living on less than $800 per year. Spread over that number, the aid total works out to an allotment of $20 per year, each. That's just not enough to do the job.

But the conservatives are also right, because as inadequate as it might be in any case, the large majority of the $60 billion per year is tossed into a bottomless latrine of corruption. This is not surprising, as the list of governments comprising the world's top recipients of foreign aid is practically a match for the list of the world's most corrupt regimes. Basically, the rulers of these countries are little more than crooks who do not give a fig about their people, so naturally they see foreign aid as an opportunity for loot. In one of many examples that Easterly cites, during a famine in the 1990s the warlords of Somalia did everything they could to worsen the general starvation, so as to increase the amount of aid available for theft.[5] He then summarizes his case nicely, stating: "Foreign aid amounts to transferring money from the best governments in the world to the worst governments in the world."[6] To lessen local graft, therefore, donor countries frequently demand that a certain fraction of their foreign aid largesse be spent by the recipients on contracts awarded to companies from the donor country itself. In that case, it is politically connected corporations and/or high-priced consultants from the professional "foreign aid community" who get the dough. Americans witnessed a mild version of this process in their own country in the case of the aid effort after Hurricane Katrina. Congress appropriated $62 billion in aid, enough to give $100,000 outright to each man, woman, and child of the roughly six hundred thousand displaced victims.[7] The money vanished instantly into the pockets of insider contractors, with only a small fraction making it to those in need.

Of course, the officials of the foreign aid agencies like to have visible structures built with their money, so there is something tangible they can point to when asking their home country politicians for next year's budget. This prioritization fits the needs of the Halliburtons of the world as well. So roads and schools do get built in some numbers. But after the ribbon-cutting ceremonies are over and the aid officials fly home, the last thing the local kleptocrats have in mind is to waste their own share of the take paying road maintenance workers or teachers. No, such funds are needed to fill their numbered accounts in Switzerland. After all, they have their families' future well-being to consider, and no good thing ever lasts forever.

The private development organizations tend to spend their tiny fraction of the total aid budget rather more effectively, directing it toward focused local projects. But after Oxfam or WaterAid have done their work and provided a village with its first clean well, what remains is an impoverished village, with good water perhaps, but no work, no pay, and little or nothing to eat. (Unfortunately, if an aid organization should step in to resolve the latter problem by distributing food, the local farmers would be ruined.)

Despite the overall failure of aid efforts, a few formerly impoverished nations have advanced substantially over the past fifty years, so much so that some of them are no longer considered to be part of the "third world." These include India, China, Taiwan, South Korea, and Singapore. It is an interesting and important fact, however, that all of these nations accomplished this transition without Western development aid playing a substantial role. Instead, following the earlier example of Japan, they earned their way to prosperity by finding something to sell.

There is a lesson here. Charity cannot and will not develop the third world. If we want to help lift the underdeveloped sector out of poverty, there is only one way to do it. Instead of giving them our aid, we need to give them our business.

Give a man a fish, and you will bankrupt his neighbor, the fisherman. Offer to buy a lot of fish, and the fisherman will hire the man to help him to meet your order. Then the man will have money, and he'll be able to buy whatever he wants, instead of being stuck with nothing but your wretched leftover fish.

DEVELOPMENT THROUGH TRADE

> Reform of agricultural trade is of central importance for many developing countries and is an essential ingredient of the negotiation and its outcome.
>
> —Joint statement regarding the Doha Round Trade Talks of the governments of Argentina, Bolivia, Botswana, Brazil, Chile, China, Colombia, Cuba, Dominican Republic, Ecuador, El Salvador, Gabon, Guatemala, Honduras, India, Malaysia, Mexico, Morocco, Nicaragua, Pakistan, Paraguay, Peru, Thailand, Uruguay, Venezuela, and Zimbabwe, June 6, 2003

Following World War II, leading economic thinkers from the victor nations came together to reconstruct the global business framework. Central to the new design were provisions for free trade. Protectionist trade battles had greatly worsened the depression, which in turn had led to fascism and war. International agreements needed to be put in place to prevent another such disaster.

Accordingly, a multilateral treaty was signed in Geneva in 1948 to establish the General Agreement on Tariffs and Trade (GATT). The GATT became a charter governing most of international trade until

1995, when it was replaced by the World Trade Organization (WTO). In many ways the GATT was quite successful. It liberalized trade in manufactured goods substantially, reducing tariffs and other trade impediments, thereby contributing significantly to the rapid reconstruction and economic growth of the advanced sector countries in the postwar period. Taking advantage of the opportunities the GATT opened to growth driven by manufactured exports, Japan multiplied its per capita income eight times over between 1948 and 1973, and doubled it again by 1995.[8] Taiwan, South Korea, and then other east and south Asian nations with industrial potential and outward-oriented economic policies were able to follow its example.

But the GATT was a system run by the industrialized nations, for the industrialized nations (of the rich, by the rich, for the rich), and its successor, the WTO, has continued that tradition.[9] For this reason, the GATT and WTO have focused their efforts on liberalizing trade in commodities that industrial nations export, while scrupulously avoiding most attempts to open advanced-sector markets to agricultural goods produced in the third world. As a result, reports Nobel Prize–winning economist Joseph Stiglitz, "the average OECD tariff on imports from developing countries is four times higher than that on imports originating in the OECD."[10]

The OECD, or Organization for Economic Cooperation and Development, is the club of the world's richest and most economically advanced nations. Effectively, the GATT/WTO rules say that if you are not in the club, you'll have to cut your prices to the bone if you want to sell anything to its members. The poor are thus forced to accept less than the rich in exchange for the same product.

This unfair system has caused the world's worse-off nations incredible economic damage. As former US president Jimmy Carter noted in Johannesburg during the UN Summit on Sustainable Development in September 2002, "We cost the developing world three times as much in trade source restrictions as all the overseas development assistance they receive from all sources."

The effects of a tax of $180 billion per year on the world's poorest people are little short of genocidal. The WTO has therefore become the target of increasingly vociferous complaints from the developing nations, the Catholic Church, nongovernmental organizations such as Christian Aid and Oxfam, and, rather less coherently, large mobs of enraged street demonstrators in Seattle (1999), Melbourne, Prague, Washington (2000), Genoa (2001), and many cities since. In order to address the concerns raised from these varied quarters, the WTO called a meeting in Doha, Qatar, in November 2001 that launched a new "Development Round" of world trade talks.

The Doha Ministerial Declaration, adopted on November 14, 2001, made a bold humanistic statement of the goals of the new negotiations: "International trade can play a major role in the promotion of economic development and the alleviation of poverty. We recognize the need for all our peoples to benefit from the increased opportunities and welfare gains that the multilateral trading system generates. The majority of WTO members are developing countries. We seek to place their needs and interests at the heart of the Work Program adopted in this Declaration."

Commenting on the declaration, US trade representative Robert Zoellick was quite clear on exactly what trade barriers needed to be brought down: "Doha lays the groundwork for a trade liberalization agenda that will be a starting point for greater development, growth, opportunity, and openness around the world. . . . [W]e've settled on a program that lays out ambitious objectives for future negotiations on the liberalization of the agriculture market. These objectives . . . will help the United States and others to advance a fundamental agricultural reform agenda."[11]

But it didn't happen. Far from it: Instead of making any concessions on the issue of their agricultural trade barriers, the United States, the European Union (EU), and Japan walked into the next WTO meeting in Singapore in 2002 with a new set of demands centering on their own issues, primarily concerning international protection of intellectual property.

Now it is quite true that many third world countries are notorious in allowing pirated editions of books, videos, software, pharmaceuticals, and other copyrighted or patented items to be produced and marketed within their borders, and this practice has incurred significant losses to advanced-sector corporations. Yet when all is said and done, the issue of piracy's impact on Hollywood's or the pharmaceutical giants' global distribution profits is rather inane when advanced to change the subject from Doha's original agenda of opening American, European, and Japanese markets to desperate third world farmers. The underdeveloped nation delegates were understandably frustrated.

In May 2002 the US Congress made things even worse by increasing domestic farm subsidies by $83 billion over the next ten years (to $190 billion total). The EU followed suit in early 2003, approving a 56 percent subsidy for European crops. While reducing costs to domestic consumers (at the expense of domestic taxpayers) for food products, such subsidies are just as effective as tariffs in blocking imports, since they force a foreign supplier to drop his price to match the subsidized level.

The crisis came the following year at the WTO meeting in Cancún, Mexico. Despite repeated pleas and a forceful joint statement on June 6, 2003, from the twenty-seven most important Latin American, African, and Asian nations, the OECD countries wouldn't give an inch. Finally, on September 15, 2003, after three months of fruitless talks, the Kenyan minister dramatically walked out of the hall, to be followed within hours by most of the rest of the third world delegates. Seeing no alternative, the meeting's host, Mexican finance minister Luis Derbez, declared it closed that very evening.

A few more attempts were made to revive the negotiations, and the Doha Round Talks limped on from meeting to meeting for three more years before they finally collapsed in total failure in June 2006.

Doha was a disaster of historic proportions. International trade negotiations are an obscure subject, so Doha will never be as famous as, say, the diplomatic catastrophe of August 1914 that led to World War I. But it should be. As a consequence of the failure of the Doha Round, tens of millions of people will be denied their livelihood—and millions of them will die.

Why did Doha stall? It was not because of lack of humanitarian feeling. On the contrary, the political classes and government bureaucracies of the OECD nations are filled with people overflowing with humanitarian feelings. They have all the caring, sympathetic, humanitarian feelings anyone could ever want from such people, and more. But what they don't have is a market in their home countries capable of absorbing third world agricultural produce without bankrupting their own nations' farmers. Until they do, the economic holocaust will continue.

That is why we need to move to the alcohol economy.

ALCOHOL FUELS AND WORLD TRADE

Recall: Ethanol is made from food crops with a large sugar or starch content. Methanol can be made from any kind of biomass, as well as from coal, natural gas, and recycled wastes. How large a market for the world's agricultural sector could be created if we switched our cars to flex-fuels?

As I write this, the price of oil is hovering around $70 per barrel, and 76 million barrels a day are being sold worldwide. That works out to a revenue stream of $5.5 billion per day, or $2 trillion per year. About 40 percent of the total is produced by OPEC. Let's say our goal is to wipe OPEC out by replacing all of their oil with alcohol. This would be an excellent move, since doing so would essentially annihilate the financial base of Islamofascism, not to mention the Iranian nuclear bomb program. In the process, however, we would also liberate $800 billion per year, much of which could go to agriculture.

To get an idea of what this would mean, take a look at table 8.1, where I present a list of the world's thirty top agricultural nations along with the value of their 2005 crops.[12]

Sent into the world agricultural sector, $800 billion per year would represent a global farm income boost of nearly 50 percent. But that is total income; most farm products are consumed domestically. Relative to international agricultural trade, the impact would be much larger.

TABLE 8.1. TOP THIRTY AGRICULTURAL NATIONS

Country/Territory	Yield ($millions)
European Union	292,820
China	263,952
India	151,533
United States	127,700
Japan	64,415
Brazil	60,560
Turkey	40,341
Indonesia	37,040
Russia	37,035
Mexico	28,680
South Korea	27,607
Australia	25,996
Canada	23,034
Iran	21,535
Nigeria	20,925
Pakistan	19,915
Argentina	19,110
Thailand	16,823
Philippines	13,364
Bangladesh	13,294
Colombia	12,612
Ukraine	12,515
Egypt	12,220
Morocco	11,435
Vietnam	9,578
Saudi Arabia	9,279
Malaysia	8,935
Sudan	8,908
Algeria	8,886
World	1,756,800

To see this, let's consider the case of the United States. The United States imports 12 million barrels of oil per day, more than a third of the OPEC total. In 2006 this oil cost us about $260 billion. If we were to spend this on alcohol fuels instead, we could increase the *total* US farm income of $127 billion by 50 percent—and still have around $200 billion left over to pay for alcohol fuel imports derived from third world agriculture. For comparison, US agricultural imports in 2005 totaled $56 billion. Thus, by switching to alcohol, we could *quadruple* our purchases of third world agricultural goods, while giving US farmers substantially *more* business, not less.

America, as mentioned, is responsible for about one-third of global oil imports. If the pattern described above were followed by other oil importers as well, the net effect would be to multiply total third world agricultural foreign exchange earnings fourfold, yielding earnings that would dwarf all current worldwide foreign aid expenditures by a factor of ten. Moreover, while most foreign aid given to third world governments is stolen by the gangsters who rule them, in this case the money would go to the productive sectors of the poor nations' economies. The kleptocrats would seek to tax this income, of course, but their ability to do so would be severely constrained by the need to keep their country's alcohol exports competitively priced. A huge engine for world development would thus be created.

The above analysis is oversimplified in a number of ways. Methanol can be produced from sources other than biomass. So in the absence of tax or tariff policies favoring renewable options for alcohol fuel production, not all of the funds transferred away from the oil cartel would go to agriculture. Furthermore, OPEC nations would no doubt attempt to counter a shift to alcohol fuels such as that described above by dropping their prices. This in itself might be considered a major achievement on behalf of humanity for the flex-fuel mandate policy. Both advanced and third world nations would benefit from the reduced oil price, and the latter could still obtain a critical improvement in their foreign exchange position as well as increased farm income through the substitution of home-grown alcohol for imported petroleum. However, if the goal is to defeat the OPEC terror patrons rather than just constrain the growth of their income, further measures would be required. These could range from punitive tariffs, counter-embargoes, and alcohol-biased fuel tax policies, to blockades or air strikes on oil export facilities in need of more rapid termination. We shall return to this subject in chapter 12. Suffice at this point to say that, were it in our interests to do so, we could shut down the oil exports of such major terror sponsors as Iran and Saudi Arabia at our discretion.

In any case, what is clear is that the potential market for agricultural products made possible by transitioning from the petroleum economy to the alcohol economy is huge—much greater than would have been required to break the impasse at Doha, and large enough, in fact, to uplift the world.

CROPS FOR THE ALCOHOL HARVEST

In the United States a strong young ethanol industry is rapidly growing based upon ethanol produced from corn. In Brazil sugarcane is already serving as the base for an ethanol fuel industry so large that it has made the country energy independent. But these are by no means the only crops that could serve as ethanol resources. Other attractive ethanol crops include sweet sorghum, sugar beets, fodder beets, wheat, barley, rye, rice, sorghum, Irish potatoes, cassava, sweet potatoes, yams, taro, grapes, bananas, nipa palm, and sago palm.

In table 8.2 I provide estimates of the potential ethanol yields of many of these crops.

There are about 4 liters to a gallon, and 2.5 acres per hectare, so 10 L/ha is nearly the same as 1 gallon per acre (i.e., to get gallons per acre from the ethanol in L/ha results reported in table 8.2, just divide by 10). In farming, yields per acre can vary considerably, so the quantities cited in table 8.2 should be considered as typical, rather than exact, and considerable improvements are possible. For example, in a

TABLE 8.2. ETHANOL YIELDS FROM CANDIDATE CROPS[13]

Crop	Crop Yield (ton/ha/yr)	Ethanol (liters/ton)	Ethanol (liters/ha/yr)
Sugarcane	90	90	8,100
Sweet sorghum	80	80	6,400
Sugar beet	50	90	4,500
Fodder beet	90	90	8,100
Wheat	2.7	400	1,080
Barley	3.3	410	1,270
Oats	2.2	410	902
Rice	3.5	480	1,680
Corn	9.4	440	4,136
Sorghum	3.5	440	1,540
Potatoes	44	110	4,840
Cassava	30	170	5,100
Sweet potatoes	27	167	4,509
Yams	70	160	11,200
Grapes	25	130	3,250
Nipa palm			8,000
Sago palm			1,350

recent book, one author[14] cites French yields of 7,140 L/ha/yr for ethanol from sugar beets and 2,770 L/ha/yr for ethanol from wheat, both of which are significantly higher than the table estimates. The high-ethanol-yield French wheat results are especially encouraging, because they are suggestive that with a bit more work, high-starch varieties of cereals can become economic sources of ethanol.

In any case, what is very apparent from table 8.2 is the wide variety of candidate crops, growable in every climate and corner of the world, that offer potential for ethanol production that is equal to or better than that already demonstrated by corn or even sugarcane.

One new ethanol crop that stands out as offering particular potential is cassava (*Manihot esculenta*). Cassava was originally a New World plant, but its cultivation has been taken up in many tropical countries around the world. Hundreds of millions of people in third world countries derive their sustenance from its starch-filled roots. Yields of 20 tons/ha/yr can be readily achieved in poor soil without irrigation or fertilizer, and 100 tons/ha/yr yields are possible under more optimal conditions. Before harvesting, cassava can be stored by being left in the ground, where it is resistant to pests. After harvesting, it can be stored by drying. Its foliage makes good animal feed. Taken together, these characteristics make cassava an excellent crop for small farmers in underdeveloped countries. The one downside of cassava is that, unlike sugarcane, the plant does not provide much in the way of extra biomass to use as fuel to provide the heat required to drive the alcohol production process. So an additional low-cost heat source, such as cellulosic biomass, stranded natural gas, coal, or, possibly in the future, solar mirror concentrators, needs to be used. This is hardly a deal-breaker, however, and a cassava-to-ethanol industry has begun to develop in Nigeria, with early reported yields of 410 gal/acre/yr (4,100 L/ha/yr) that already match those of American corn.[15]

Sweet potatoes and yams are also very high-yield starch (and thus ethanol) crops. They require more water and better soil than cassava, but where those are available, they offer great promise.[16]

The nipa palm (*Nipa fruticans*) is also worthy of note. This three- to four-meter-tall tree is widespread in the Indonesian archipelago, Southeast Asia, and the Pacific. It produces copious amounts of a sap that is 15 percent sugar by weight. On well-planted farms, 30,000 liters of this sap can be obtained from each hectare every year, enough to produce 8,000 liters of ethanol. This is nearly twice the current yield per hectare of ethanol from American corn. Nipa palm thus offers the potential for energy orchards that produce fuel year after year without the need for repeated plantings.

Finally, it should be mentioned that there are large quantities of potential ethanol source crops grown in the tropics every year that are currently discarded because they cannot be preserved long enough for delivery to market. It is estimated, for example, that in Ecuador alone, enough spoiled bananas are wasted in this way every year to produce 68,000 tons of fermentable carbohydrate.[17] These spoiled fruits could be converted to ethanol.

Some plants also yield oils that can be used to produce long-chain hydrocarbons suitable for use as diesel fuel. Two of the most promising biodiesel plants are oil palm and coconut, with reported yields of 5,080 and 2,300 L/ha/yr, respectively.[18] Rapeseeds, peanuts, sunflowers, and soybeans also offer promise as sources for biodiesel, although their current yields are still in the 1,000 L/ha/yr range.

Unlike ethanol, biodiesel in general has yet to achieve sufficiently favorable economies for a major takeoff. This is shown in figure 8.1, which compares global ethanol production with biodiesel.[19] However, as we shall discuss in the focus section at the end of this chapter, there

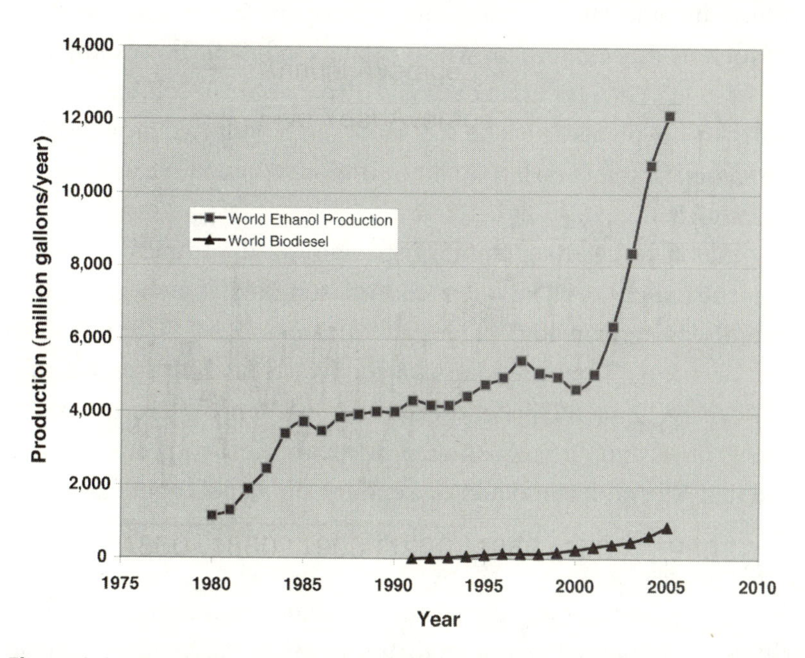

Figure 8.1. Worldwide ethanol and biodiesel production. ***Source*****: Data from F. O. Licht, Worldwatch, compiled by Earth Policy Institute, http://www.earth-policy.org.**

is an alternative path to large-scale renewable production of diesel fuel based upon methanol, which is also the simplest fuel to produce for internal combustion engines as well.

The potential crop base for producing methanol includes all plants, without exception. Not only the edible and inedible parts of commercial crops, but weeds, wild jungle underbrush, trees, grasses, fallen leaves and branches, water lilies, swamp and river plants, seaweed, and algae, can all be used to produce methanol.

The resource base such sources represent is enormous. For example, it has been calculated that about 4 billion tonnes of agricultural crop residues are produced worldwide every year, with a total energy content equivalent to 1.5 billion tonnes of oil per year, or 28 million barrels per day.[20] If it were all converted to methanol, this source alone would be enough to replace the combined oil exports of all of OPEC. Furthermore, since agricultural residue comes from farms, its utilization to produce methanol will provide farmers with an additional income stream that will improve the economics of other crops, such as those being grown to produce ethanol. For example, a farmer growing corn for ethanol would also be able to sell the vegetative parts of his plants to make methanol, and, with two crops to sell for the same effort and cost, find himself more competitive and able to sell more of both.

But that is just the beginning. Especially in the tropics, there are many plants, such as eucalyptus, cottonwood, and switch grass, that can readily be grown and harvested in amounts of 50 tonnes per hectare per year. This is enough to produce 15 tonnes, or roughly 20,000 L/ha/year of methanol, equivalent in energy value to 15,000 L/ha/year of ethanol. The Australian national science agency has calculated that Australia could produce enough methanol to replace all its oil by planting fast-growing trees on an area equal to a square 580 km on a side.[21]

By making use of tougher varieties of plants that are not ordinarily considered commercial crops, methanol farming can be done on land that is too poor or too dry for conventional agriculture. For example, the mesquite tree grows widely in semidesert regions in the American Southwest without any irrigation whatsoever. It accomplishes this by making use of an extensive root system, which also holds the soil in place. In addition, the mesquite's roots carry bacterial nodules that fix nitrogen in the soil, fertilizing it to the benefit of both the mesquite and other neighboring plants. Planted across sub-Saharan Africa, mesquite orchards grown for methanol production could do much to stop the spreading desertification while providing the region with a desperately needed cash crop.

Aquatic plants also offer tremendous potential for methanol production. For example, the water hyacinth is a free-floating plant that forms dense rafts that clog rivers, canals, and lakes all over South America and Africa, as well as many bodies of water in the southern United States. Currently, it is just a major nuisance that has to be repeatedly cleaned from such waterways to make them navigable. But in a methanol fuel economy, the water hyacinth's high productivity of 30 to 80 tonnes dry mass per hectare could turn this pest into a prime commercial resource. Another promising aquatic methanol crop is cattail, which grows without help in swamps everywhere, with typical productivities on the order of 40 tonnes per hectare. Kelp grows naturally in bays and other protected bodies of salt water, and can be farmed to bring yields up to 30 tonnes of biomass per hectare. In places where there is some industry and plentiful sunlight, carbon dioxide emissions from the industrial facilities could be bubbled into natural or artificial ponds and converted into biomass by microscopic algae. Such algae could be continually harvested, dried, and turned into methanol. Production rates for such systems could be enormous, limited primarily by the availability of carbon dioxide waste to transform.

Other sources of biomass for methanol production include materials such as sawdust and rice bran that are ordinarily viewed as waste. One such substance that is particularly attractive for methanol production is black liquor, a noxious by-product of the paper industry. Worldwide, 170 million tonnes of this pollutant are produced each year. Turning it into methanol would not only eliminate it as an environmentalist contaminant, but produce enough fuel to replace the oil exports of Iran.

While not an agricultural product, an important potential third world resource that now largely goes to waste is stranded natural gas. Natural gas produced in the United States and other advanced countries can generally (but not always) be brought to market economically because there are pipeline systems in place to transport the gas. However, in third world countries, which lack pipeline systems and are distant geographically from advanced-sector consumer markets, the natural gas released from the earth as a by-product of oil exploration cannot be effectively delivered and is therefore often flared (i.e., burned off). This is also frequently done in offshore drilling rigs. Worldwide, about 3.8 trillion cubic feet of natural gas is flared every year, while another 11 trillion cubic feet are pumped back underground.[22] If all of this were converted to methanol, it could be transported and provide 650 million tonnes of liquid fuel per year—enough to replace that exported by Saudi Arabia.

In all cases, the basic method of producing methanol from biomass is the essentially the same as that needed to make it from coal, natural gas, or municipal waste: First turn the biomass into synthesis gas, and then react the syngas to make methanol. In certain respects, biomass is better feedstock for making syngas than other sources, because biomass contains much less sulfur and other unwanted containments. However, one special difficulty faced by biomass methanol production is the need to transport the relatively bulky and heavy (for its value) biomass material to the methanol-processing plant. This problem can be dealt with in two ways.

One idea is to mass-produce small truck-mounted methanol-processing units and disperse them out into the country close to the sources of production. This eliminates the need to transport heavy crop materials long distances. While chemical plants can ordinarily be constructed for lower cost per unit of output by taking advantage of economies of scale, this approach would employ the technique of mass production of identical units in order to keep the price of small systems down. It would also have the advantage of enabling a large number of small players to get into the manufacturing side of the methanol production business, thereby promoting the growth of a third world middle class.

Alternatively, if it is desired to make use of large-scale, multithousand-ton-output-per-day methanol production facilities, then mass-produced small-scale units can be sent into the country to reduce the size and weight of the crops prior to shipment by employing a process called flash pyrolysis. In flash pyrolysis, biomass particles are heated very rapidly to 500°C to produce a gas mixture that is quickly quenched, resulting in the formation of a black liquid called biocrude. Biocrude can actually be processed like oil to produce petroleum-like products. Alternatively, it can simply serve as a compact intermediate substance to enable economic transport of the biomass from the countryside where it is grown to a centrally located methanol production plant.

In both cases, it should be noted that the alcohol economy will also create nonfarm industrial jobs in the countryside, and thus a mechanically adept blue-collar workforce, which is another important necessary element for economic development.

Finally, there is coal, which is plentiful in the United States, Europe, India, China, Australia, South Africa, and the former Soviet Union. Coal is cheap (~$0.03 per kg), and, if necessary, the United States could make all the methanol it needs from domestic coal. The fact that the alcohol economy gives us this option defines it as the critical solution to a central problem of US national security.

But we actually have a labor shortage in America right now, as shown by the fact that we need to bring in millions of immigrants to do the work available. So rather than exercise our alcohol option for energy autarchy, it would behoove us to create jobs south of the border by buying much of our methanol and ethanol abroad. Because, unlike oil, the sources of alcohol are so diverse, we can allow ourselves to count upon importing a significant fraction of our supply without impacting our safety. The advantage of doing this is that by allowing the tropical agrarian countries to produce some of our alcohol, we will put money in their hands that they can use to buy our manufactured goods, such as tractors and harvesters to grow the fuel crops, trucks to transport them, and equipment to process the biomass into liquid fuel.

It's called fair trade, and it benefits everyone. The hydrogen economy offers no possibility for enhancing world trade, because hydrogen cannot be transported economically. But alcohol can. By opening up the world trading system, the alcohol economy offers the prospect a new age of global development.

Half the world's people need not remain imprisoned in poverty: their aspirations crushed, their creative gifts strangled. They can escape. To do so, they don't need charity, expert advice, or condescending coercion. What they need is a piece of the action. The alcohol economy will make that possible.

FOCUS SECTION: ALCOHOL ALCHEMY

OPEC currently holds the world in a stranglehold because it controls the source of our liquid fuels. As we have seen, we can break this critical dependence by moving to the alcohol economy. However, petrochemicals are currently used not just to make fuel, but to produce plastics, synthetic fibers, and many other materials that, since their advent in the mid-twentieth century, have become vital to our civilization. If we are to truly destroy the financial power of the terror bankers, we must find ways to eliminate the need for oil to serve these purposes as well. Fortunately, alcohols provide us with the means to do so.

The key starting point for the alternative synthetic chemistry is methanol. As Nobel Prize–winning chemist George Olah has shown,[23] nearly every synthetic plastic currently made from petroleum products can be efficiently produced through methanol chemistry instead.

The key processes begin by reacting methanol with itself to form dimethyl ether via the reaction below.

$$2CH_3OH \rightarrow (CH_3)_2O + H_2O \quad (1)$$

Reaction (1) is only marginally exothermic, but any extra energy needed to drive it can be obtained by burning a little bit of the methanol. It occurs quickly at 400°C and 1 atmosphere pressure using cheap gamma alumina pellets as a catalyst.

As remarked in chapter 6, the dimethyl ether produced in this manner makes excellent diesel fuel. Dimethyl ether burns without producing soot, black smoke, or other particulates, and contains none of the sulfur or other impurities that pollute the exhaust of conventional diesel fuel. In addition, with a cetane rating of 60 (compared to about 48 in petroleum-derived diesel fuel), DME offers significantly better performance. Because it is made from methanol, there is virtually no limit to the amount of dimethyl ether we can produce. Its energy density is only about half that of regular diesel, so fuel tanks would have to be larger for the same range, and they would have to be somewhat stronger too, because DME stores at a moderate pressure of about 75 psi (5 atmospheres) at room temperature. These latter two features are burdens, but not particularly onerous ones.[24] The ultimate test that will determine DME acceptance, as with most fuels, is cost. It takes about 1.4 kg of methanol to make 1 kg of DME. Methanol can be made for $0.93 per gallon, or $0.30 per kg. At this price, large-scale production of DME from methanol should be possible for about $0.45 per kg, or $1.20 per gallon. This equates on an energy basis to conventional diesel fuel at $2.40 per gallon, which is about 20 percent less than the going rate.

Thus, in an alcohol economy, mass production of DME from methanol could provide an economical, renewable source of nonpolluting fuel for trucks, trains, and ships. With its moderate vapor pressure, DME also makes a good replacement for propane for use in portable cooking systems and other domestic applications.

Outside of the US government, these possibilities have not gone unnoticed. China, in particular, seems to be moving to embrace DME, with the construction of a plant in Shandong Province that will produce 1 million tonnes of DME and 1.5 million tonnes of methanol per year now under way. The Japanese Ministry of Economy, Trade, and

Industry has made a commitment to the development of mass-produced lost-cost DME, and several large Japanese consortia, led by the NKK Corporation and Mitsubishi Gas Chemical, have announced plans to build DME plants in the 1- to 3-million-tonne-per-year range. Other countries that have expressed serious interest in developing DME fuel production facilities include India, Australia, Iran, and Trinidad and Tobago.

However, DME is not just a fuel; it is a chemical building block that can be used to make many other things. For example, by heating DME at 1 atmosphere pressure to about 400°C in a reactor filled with the zeolite catalyst ZSM-5, ethylene (C_2H_4) can be readily produced. The reaction is:

$$(CH_3)_2O \rightarrow C_2H_4 + H_2O \qquad (2)$$

Reaction (2) is mildly exothermic, yielding 7.4 kcal/mole, so once you get it started, it can provide most of the heat to keep itself going. If you raise the temperature to about 450°C and the pressure to 1.5 atmospheres, instead of ethylene, propylene (C_3H_6) can be produced. The same zeolite-catalyzed reactor can be used to make both, with output ratios governed by the pressure and temperature conditions. Thus, a plant using this system could alter its output of ethylene and propylene to take advantage of changing market conditions. Using similar chemistry, ethanol can also be used to make ethylene.

Now, ethylene and propylene are the two most important feed-stocks for the modern plastics industry. When heated under pressure in the presence of appropriate catalysts, they can be polymerized to make polyethylene and polypropylene, respectively. These two materials, by themselves, represent over half of the entire worldwide plastic market. Polyethylene is used to make all kinds of cheap items, from toys to food storage containers to garbage bags. Polypropylene is somewhat more expensive, and is used to make higher-quality plastic items as well as excellent synthetic fabrics.

Propylene, however, is a very versatile molecule, and if compressed to 20 atmospheres in a ZSM-5 reactor at 400°C can be reacted with itself to build higher hydrocarbons, including straight-chain alkanes such as butane, pentane, and hexane, as well as "aromatic" ring molecules such as benzene, toluene, and various xylenes. In addition to being important in their own right, and as ingredients for additional plastics such as polystyrene (the source of our society's now omnipresent Styrofoam), these chemicals can also be used to make synthetic gasoline. If longer carbon chains are needed to make heavy fuels, that can be done, too: Just substitute tungsten oxide on alumina for the reactor catalyst instead of ZSM-5.

These chemical sequences may seem involved, but they are quite well understood. They were demonstrated to be practical on a significant scale in the early 1980s, when, reacting to the second oil shock, the government of New Zealand decided to put them into practice. A plant was built to take natural gas from the large offshore Maui field and turn it into methanol, after which the methanol was turned into high-octane synthetic gasoline. Total output was about 600,000 tonnes per year, which was enough to provide about one-third of the entire nation's gasoline supply at that time. The plant worked quite well. Unfortunately, however, when OPEC dropped its oil price to $10 per barrel in 1986, the government did nothing to protect its energy independence initiative, and all synthetic gasoline production at the facility stopped. The plant still makes methanol, however, for sale on the world chemical market. Perhaps now that the price of petroleum has skyrocketed, the New Zealanders will reconsider their decision to depend on the oil cartel and put the gasoline production facility back into action.

These are just a few examples of the kind of chemical engineering tricks you can play with methanol and ethanol. The point, however, is this: If you can make it with petroleum, you can make it with alcohol. And that is very good news for a world that needs to escape the bonds of oil.

CHAPTER 9

THE BRAZILIAN EXPERIENCE

In many ways, Brazil is a microcosm of the world. Its diverse, multiracial society includes wealthy sophisticates with educations, ideas, and living styles mirroring those of New York or Paris; unionized blue-collar industrial workers not too different from those of Detroit or the German Ruhr Valley; and urban slum dwellers, rural peasants, and downtrodden aborigines whose poverty and desperation fully match their counterparts throughout the third world. The nation's annual gross domestic product (GDP) per capita is $8,100, almost equal to the world average of $8,800, and its governmental system has fluctuated between various combinations of democracy, oligarchy, kleptocracy, technocracy, and military dictatorship.

Brazil is a big country, with a grand vision of its own destiny, and its leaders, for good or ill, have shown a penchant for implementing dramatic policies with bold strokes. In 1888 Princess Isabel, taking advantage of her temporary power while her father, Emperor Pedro II, was away from the country on a European trip, issued a proclamation emancipating, without compensation, the slaves who comprised a third of the empire's population. This move so infuriated the conservative landowners who provided the monarchy with its political base that it cost the imperial family its throne. But the decree stuck, and Brazil rose from slavery.[1]

In 1956 populist president Juscelino Kubitschek decided that the time had finally come for Brazilians to move on from the Atlantic shore and actually settle the interior of their vast country. The way to do it, he thought, would be to move the nation's capital from coastal Rio de Janeiro to an inland city. Unfortunately, there were no existing inland cities worth considering. So he decided to build one from scratch, on a barren plain in the middle of nowhere, some six hundred miles from the sea. Furthermore, not only would the city be new, it would be a monument to modernity, with buildings, sculptures, and an overall layout reflecting the thinking of some of the most visionary architects and artists in the world. To most outsiders, the project seemed utterly fantastical, and yet, at great cost and heroic effort, Brasilia was built—*and commissioned within Kubitschek's five-year term.* Offering the highest quality of life in the country, it has since grown to a metropolis of 2.3 million people and has powerfully served its original intended purpose of acting as a driver for the development of the nation's interior.

So it was following these traditions that in 1974, when the newly inaugurated General Ernesto Geisel found his nation confronted with economic devastation from the Arab oil shock, he acted forcefully to seize his moment for greatness. Even though Brazil imported 80 percent of its oil, the enlightened military autocrat decreed that it could and would become energy independent. His answer was ethanol.

The choice was not fortuitous. Brazil has been the world's leading sugar producer since the seventeenth century, and has used sugar to make alcohol (via molasses to rum) since that time. Research on making use of ethanol to power automobiles was started in Brazil as early as the 1920s. During World War II Brazil joined the Allies, and in the face of German U-boat interference with its oil imports, was forced to make use of sugar alcohol to stretch its fuel supplies.

Prior to his presidential assignment by the country's ruling military oligarchy, Geisel had run Petrobras, the Brazilian government's oil company. He was well acquainted with the nation's fuel history. Without hesitation, he issued an edict requiring that all gasoline sold in Brazil contain a 10 percent mixture of ethanol.

Geisel had big plans for Brazil. He wanted to turn it into a nation of the first rank, with nuclear power and advanced manufacturing industries. Eliminating 10 percent of the nation's oil import bill saved him critical foreign exchange to buy the technology he needed to make progress. But switching to alcohol also served another purpose: It saved the nation's agricultural sector, which had been hit hard by a devastating 72 percent drop in international sugar prices during 1975.

The country's sugar producers were delighted by their new market, but for Geisel 10 percent was not enough. He wanted to go much further; however, doing so would require radical expansion of the nation's ethanol production capacity. So he instituted programs providing low-interest soft loans for the building of hundreds of new ethanol distilleries, and as capacity rapidly rose, he ordered the ethanol mixture requirement in gasoline to be ramped up to match.

The results of the program were spectacular. Ethanol production soared 500 percent in four years, allowing the mandated alcohol blend ratio to be raised to 24 percent. Geisel wanted to achieve even more, and it was clear that the nation's booming sugar alcohol industry could provide the fuel to do so. Yet 24 percent ethanol is about the maximum that ordinary cars designed for gasoline use can accept. Without new technology, the program had reached its limits.

Enter Umberno Stumpf. A researcher at the Brazilian air force laboratory in the southeastern state of São Paulo, Stumpf approached Geisel and pitched to him that the way forward was for Brazil to produce its own cars, designed to run on ethanol alone. Geisel liked the idea and provided Stumpf the funding to build three ethanol cars (a VW Beetle, a Dodge, and a Brazilian Gurgel) and send them on a widely publicized five-thousand-mile odyssey from the air force lab to the Amazon city of Manaus in November 1976. Following this success, Stumpf convinced the São Paulo state telephone company to convert its four-hundred-car fleet to ethanol.

This was a start. But it was apparent that unless a nationwide distribution system was set up, there would be no consumer demand for cars that ran on pure ethanol. However, as an autocrat, Geisel had a solution: He simply ordered every gas station in the country to install an ethanol pump. This was done at state expense, because in Brazil the gas stations are run by the national oil company. But still, it was done, and by 1979 there were more than twenty thousand ethanol pumps ready for consumer use at stations across the nation.

Since there were virtually no ethanol cars, these pumps had no customers. The program thus seemed to many to be pure folly—but not for long. In 1979 the Iranian revolution launched the second oil shock, and gasoline prices went through the roof. Geisel and his successor, General João Baptista Figueiredo, offered large tax breaks to automobile companies to speed ethanol cars into production. The automakers responded instantly. By the end of the year, Fiat was selling ethanol cars in Brazilian showrooms. Before 1980 was out, every foreign and domestic auto company operating in Brazil had ethanol models available for purchase.

Sales took off immediately. By 1981, 30 percent of all new cars sold in Brazil were ethanol powered. By 1983 the ethanol car market share was more than 90 percent. By 1986, when the Brazilian government was transferred to civilian control, more than 5 million ethanol cars had been sold. Nearly half of the country's automobile fuel was being provided by domestic ethanol, with huge savings to both consumers and the nation's foreign exchange. More than a million jobs had been created. The program seemed like an overwhelming success.

But then, in 1986, OPEC dropped its price, and the cost of gasoline fell with it. For several years, Brazil's civilian leaders struggled to keep the nation's ethanol competitive by providing price supports, but this cost the government a lot of money and provoked the displeasure of international bankers anxious to collect on Brazil's foreign debt. Under pressure from the International Monetary Fund (IMF), Brazil abandoned its ethanol price supports in 1990. During that same year, international speculation caused the price of sugar to triple. As a result

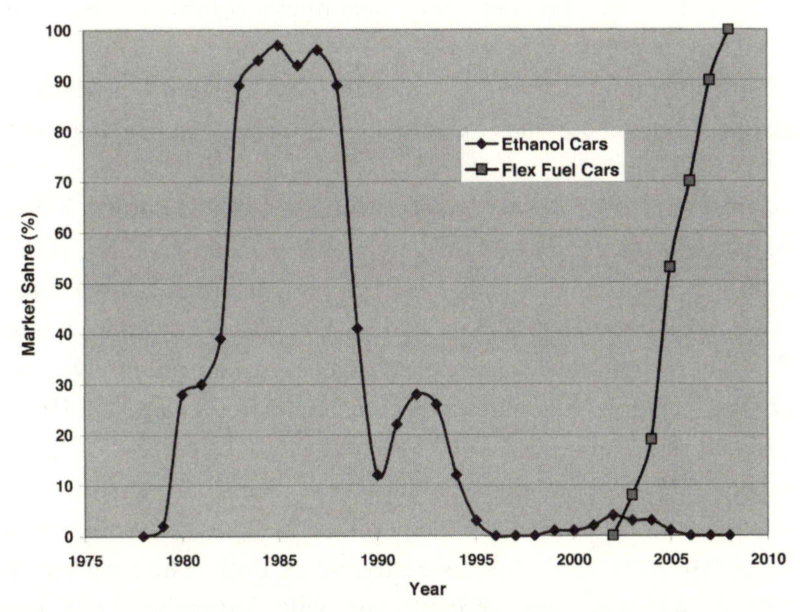

Figure 9.1. The rise and fall of the Brazilian ethanol car markets. Flex-fuel shares for 2007–2008 are projections. *Source*: Data from Alfred Szwarc, "Use of Bio-Fuels in Brazil," presentation to In-Session Workshop on Mitigation, SBSTA 21 / COP 10, Buenos Aires, December 9, 2004; and David Sandalow, "Ethanol: Lessons from Brazil," Brookings Institute white paper, 2006.

of these two factors, the price of Brazilian ethanol shot up well above the price of gasoline, shortages occurred, and the market for the ethanol cars collapsed. From 1989, when ethanol cars held 90 percent of the market, sales fell to 20 percent in 1991, and they fell further to 3 percent by 1995.

The dramatic rise and fall of the Brazilian ethanol car industry, and the subsequent rise of the flex-fuel industry, is shown in figure 9.1.[2]

To all appearances, it seemed as though the Brazilian ethanol experiment had collapsed in ignominious failure. Yet throughout the 1990s the government defied IMF pressure and kept the industry alive by maintaining the requirement for 20 percent ethanol in all regular gasoline sold. The ethanol pumps were kept operating as well, to meet the needs of the existing, but aging, fleet of ethanol cars that had been created during the prior boom period.[3]

Most important, led by the visionary physicist and Science Minister José Goldemberg, the ethanol believers within the Brazilian technocracy did not give up. Through the dark years of the 1990s, researchers at the Center for Sugarcane Technology in São Paulo kept working away identifying and developing new sugar varieties, slowly increasing ethanol yields to 6,000 liters per hectare and driving production costs down to less than a dollar a gallon. Meanwhile, licking their wounds from the ethanol car debacle, Brazilian automotive engineers eyed the flex-fuel vehicles that had begun to appear in the United States and resolved to create their own versions. Sugar and oil prices were both highly volatile; in the late seventies and early eighties they had made ethanol the cheaper fuel, only to swing in favor of gasoline later on. But they could just as easily swing back. The right car for Brazilians would be one that could deal with that reality. Such a car could only be flex-fueled.

It was Fernando Damasceno, chief engineer of the Brazilian unit of the Italian auto parts company Magneti Marelli, who made the key advance. Developed during the mid-1990s, the Damasceno system accomplished the same feat as the technology created ten years earlier by the Nichols group at Ford, but it was simpler, and thus cheaper to manufacture. Furthermore, by basing its control operation on sensing the oxygen content in the vehicle exhaust rather than the ethanol fraction in the fuel mixture, the Damasceno system was more robust and versatile, and readily compatible with a wide variety of alcohol and gasoline types.

For several years, Damasceno's invention went unnoticed, as it seemed to have no purpose so long as gasoline remained cheap. But in 1999 OPEC began jacking oil prices back up again, and people started paying attention. As gasoline prices continued to rise in 2000 and 2001, Volkswagen's Brazilian division management saw the light, and launched an effort to convince the government to grant them a tax break if they would bring the flex-fuel car to market. Obtaining this agreement, VW purchased a license to produce cars using the Magneti Marelli technology in 2002, and, not wasting a moment, came out with the flex-fuel Gol the following year.

Their timing could not have been better. Gas prices were up and the old ethanol pumps were still standing at twenty-nine thousand Petrobras stations, offering alcohol fuel at half the price. Consumers saw the new technology's advantage immediately and made their views clear with their wallets. In its first year on sale, the flex-fuel model VW Gol took nearly 10 percent of the entire Brazilian auto market.

All of the other Brazilian auto companies got the message and rushed to obtain Magneti Martelli flex-fuel licenses for themselves. In 2004, with more models on sale, the flex-fuel market share grew to 20 percent. In 2005 it was 53 percent, and in 2006 it exceeded 70 percent. As I write this, flex-fuel auto market share is projected to approach 90 percent in 2007, after which it *will* be 100 percent, since all non-flex-fuel lines are being phased out.

The success of the flex-fuel cars in Brazil may seem astonishing, but it shouldn't be. As the Brazilian public immediately realized, with flex-fuel technology they couldn't lose. They didn't have to bet in advance whether gasoline or ethanol would be cheaper, they could just choose the fuel that was. Furthermore, they could see that by owning flex-fuel cars, they could make the ethanol and gasoline vendors compete against each other for their money, and they liked the feeling. Finally, by 2005 another advantage of the flex-fuel cars became apparent: They retained more of their resale value than single-fuel vehicles. People who had earlier bought gasoline cars found themselves at a loss trying to unload them in the face of the high gas prices of mid-decade. But with a flex-fuel car that could never happen. It was simply a better mousetrap. There was no reason to buy anything else.

As a result of the success of the flex-fuel car, ethanol use rose again to provide 40 percent of all auto fuel used in Brazil by 2006. Coupled with a program of aggressive state sponsorship of offshore oil exploration, this has allowed Brazil to become energy independent.

This is shown dramatically in figure 9.2, which compares the his-

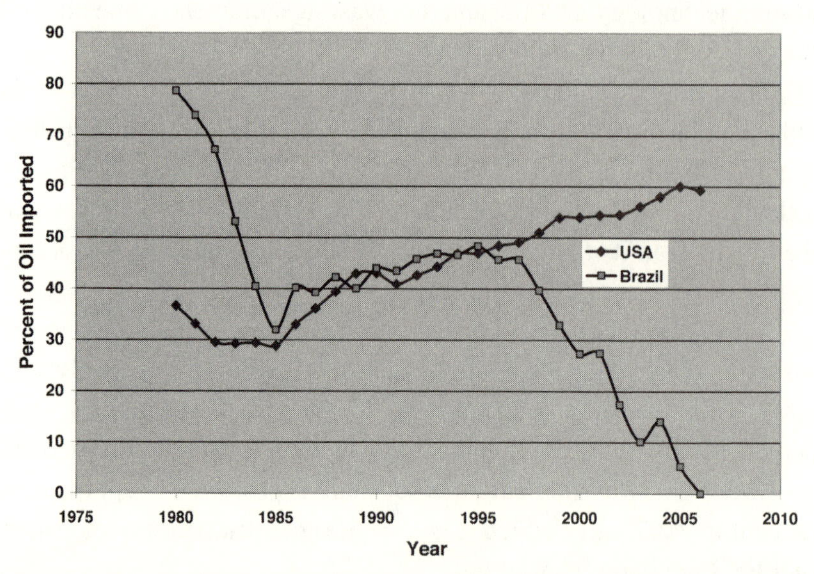

Figure 9.2. Oil dependence of Brazil and the United States, 1980–2006. ***Source*****: Data Energy Information Administration, http://www.eia.doe.gov.**

tory of oil dependence of Brazil with that of the United States over the period from 1980 to 2006.[4]

It is worth reflecting on this history. In 1974, at the time of the first oil shock, Brazil imported 80 percent of its oil and the United States imported 30 percent. By 1994, as a result of the Brazilian ethanol program on the one hand and the lack of any meaningful energy independence policy in the United States on the other, both countries imported 45 percent of their oil. Since 1993 America's dependence has grown from 45 percent to 60 percent, while Brazil's has declined from 45 percent to zero. Not only that, but in 2006 Brazil became a net energy exporter, independent in oil and earning some $600 million in foreign exchange by exporting ethanol to Sweden and Japan.

Brazil has half the population of the United States, and its per capita GDP is only one-fifth as much. Its economy, therefore, is roughly one-tenth the size of ours. Its 2006 ethanol production of 4 billion gallons will save the country about $14 billion in foreign exchange. Relative to the country's GDP, this is equivalent to a $140 billion foreign exchange savings for the United States. Brazil's total foreign exchange surplus runs about $30 billion per year. Without the ethanol program, this would be cut roughly in half.

Since its inception in the 1970s, the Brazilian ethanol program has cost the government a total of about $16 billion, an expenditure that made it objectionable to the IMF. But these costs have been repaid nearly tenfold by the $150 billion in savings in oil imports it has produced over the same period.

There have been other benefits as well. Brazilian distilleries burn the bagasse, or woody "cane" part of the sugarcane, to obtain the heat required to turn the sugar into ethanol. They then use the waste heat from this process to generate electricity. The amount of power so obtained is by no means insignificant. As of 2006, Brazilian distilleries collectively generate 600 MW of electricity, enough to power a large city, with another 3,400 MW potentially available in the future.

The employment benefits have also been quite substantial. Between farm and factory, the ethanol industry now provides the nation with 1.8 million jobs. This is approximately 2 percent of the entire Brazilian workforce.

There have also been noteworthy environmental benefits. Ethanol-burning cars release much less sulfur dioxide, carbon monoxide, and particulate emissions than their gasoline counterparts, and this has resulted in a radical improvement in air quality in many Brazilian cities. For example, according to a 2005 paper by Goldemberg,[5] the ethanol program has cut carbon monoxide concentrations in the air of São Paulo by a factor of six and sulfur dioxide by a factor of nine relative to their levels in 1980, while enabling the elimination of all polluting lead and MTBE additives in the Brazilian fuel supply. The use of sugar ethanol as automotive fuel in place of gasoline also eliminates 100 percent of the carbon dioxide greenhouse gases those cars would otherwise contribute to the environment—or, rather, more, since the extra electricity produced by burning bagasse also replaces fossil fuels.

The daily savings to consumers have also been substantial. For example, a University of North Carolina study cites a price comparison for a typical month, May 2004. At that time, gasoline sold retail for 2.02 R$ per liter (about $2.69 per gallon) while ethanol sold for 1.15 R$ per liter ($1.53 per gallon).[6] The rule of thumb in Brazil is that a gallon of ethanol will propel a car about 70 percent the distance of a gallon of gasoline. Taking this factor into account, the ethanol was 20 percent cheaper. Since 2004 the price of gasoline has gone up, making ethanol an even better bargain. For ordinary Brazilians, trying hard to make ends meet on an average income of $8,000 per year, such savings matter.

In the struggle to escape the bondage of petroleum, no other nation has achieved such a remarkable result. What can be learned from it?

For many in the developing world, the lesson taken from the Brazilian ethanol program is that it should be copied, and several have started to make the attempt. China, which is now the world's third-largest ethanol producer after Brazil and the United States, has already taken the step of mandating 10 percent ethanol (E10) in five of its most important provinces, and reportedly is planning to make E10 mandatory nationwide by 2008. Driven by direct funding from the finance ministry, as well as preferential tax policies, distilleries for converting corn to ethanol have been set up across China's northern, eastern, and central provinces with a total capacity of 325 million gallons per year. A giant plant with a capacity of 100 million gallons per year is now under construction in Hengshiu, Hebei Province, and is scheduled to start operation by the end of 2007. More plants are being set up in the south by the Henan Tianguan Group to make ethanol out of cassava grown in Laos.[7]

India, the world's number-four ethanol producer, has set up 120 distilleries to make ethanol from sugarcane. As of October 2006, a 5 percent ethanol mixture in gasoline (E5) was mandatory throughout India. In October 2007 the required ethanol level will be raised to E10, with E20 planned to follow.[8] Thailand, Malaysia, and Indonesia have all started major government biofuel initiatives, including not only sugarcane or cassava ethanol, but palm oil–based biodiesel programs as well.[9]

Within the advanced sector, the most important advocate of a Brazilian-style approach to energy independence has been Senator Hillary Clinton (D-NY). Clinton traveled to Brazil and observed the ethanol program in action there. Apparently she was quite impressed, because on May 23, 2006, she gave a major address to the National Press Club outlining an energy policy that mirrored, for better and for worse, some of the central points of the Brazilian program.

Specifically, Clinton called for taxing the oil companies on their windfall profits and using part of the revenues to set up a $50-billion-per-year energy development fund. The fund would have a goal of cutting US oil imports in half by 2025. As one of its key projects, it would help pay to set up hundreds of thousands of E85 gas station pumps coast to coast over the next ten years. Currently there are fewer than 1,000 E85 pumps nationwide (out of 170,000 gas stations), a fact that makes the existing American flex-fuel fleet, such as it is, ineffective as a means of reducing petroleum consumption. "We've got to take action on this pump issue or we're just spinning our wheels," Clinton said.

It will be recalled that it was Geisel's decision in the late 1970s to order all of Brazil's Petrobras filling stations to set up ethanol pumps—even though there were no cars to make use of them—that set the stage for the radical expansion of ethanol car sales a few years later. The persistence of such pumps through the turn of the century then enabled the current Brazilian flex-fuel car boom. Clinton's policy is thus modeled precisely upon Geisel's: Build the pumps, and the cars will come.

While it worked in Brazil—and could, in principle, work in America—the plan is wasteful. Critics might well ask: Why build pumps to serve cars that don't exist? The question is a valid one. In the 1970s Geisel had no choice but to do so. No one was going to buy ethanol cars until a nationwide filling station infrastructure existed to fuel them, and no ethanol pumps would be set up independently until there were ethanol cars, so a top-down dictate to install the pumps was needed to solve the problem. But today, flex-fuel cars exist, and cost no more than their monofuel counterparts to produce. If a law were passed that made it mandatory that all new cars be flex-fueled as a standard feature, within three years of enactment there would be 50 million cars on the road capable of using high-alcohol fuel, and under those conditions, the combination of opportunity for profit and competitive pressure would make the alcohol pumps rapidly appear everywhere. This would be a much better strategy than spending tens of billion of tax dollars to slowly build and maintain a nationwide network of dormant pumps with the hope of gradually influencing consumers to buy more flex-fuel cars. By leading with the flex-fuel mandate and using it to create a market to drive the development of an E85 infrastructure, the job of transitioning to alcohol would be accomplished much more quickly, and the program would not cost a penny of tax dollars. This latter feature is extremely important—not just to save money, but to ensure success. A program that costs billions every year is liable to be cut any time the political winds change. A program that simply mandates a superior technology as a standard feature of every new car, once established, is politically unassailable.

So the Brazil-cloned Clinton plan takes its steps toward achieving US energy independence in the reverse of the optimal order. Its backward illogic would waste billions, delay substantial accomplishment for years, and put its own hopes for success in serious jeopardy. Nevertheless, it must be said that mindlessly copying a successful model generally makes for much better policy than mindlessly copying a failed one—or than having no policy at all. So foggy-minded and clumsy as it might be, of all the energy policies offered thus far by American politicians of comparable power, Clinton's is the only one that rises to the level of mediocrity. Certainly, in comparison to the Bush administration's hydrogen hoax, Senator John McCain's Hassett-inspired "leave it all to the market" nonpolicy, or former vice president Al Gore's hysterical Malthusian prescriptions, her plan stands out like a dwarf towering above a horde of Lilliputians.[10]

Hopefully, either Clinton or one of her rivals will come forward with an energy plan that really gets it right. In the meantime, we at least finally have one worth fixing.

For that, we must thank the Brazilians.

FOCUS SECTION: THE ALCOHOL ECONOMY VERSUS NARCOTERRORISM

> We are making these drugs for Satan America and the Jews. If we cannot kill them with guns, we will kill them with drugs.
>
> —Hezbollah fatwah, 1985[11]

Brazil is not known as a major center for narcoterrorism. However, nearly all of its neighbors—including Colombia, Venezuela, Peru, Bolivia, Paraguay, and Uruguay—are. While most of the money supporting the spread of Islamofascist terrorism comes from the Saudi royal family, Iran, and other OPEC sources, a significant amount is derived from trade in narcotics. If we are to defeat the enemy, we must dry up this source of support as well.

The narcotics trade is a much greater national security threat than most people realize, because the people who control it are no longer simply greedy criminals trying to get rich. Rather, since the 1970s it has increasingly served as an auxiliary financial base and untraceable slush fund to support the terrorist operations of an expanding array of Islamofascist organizations, including Hezbollah, Hamas, Islamic Jihad, the Palestine Liberation Organization (PLO), al Qaeda, Abu Sayyaf, and the Taliban.[12] Until the 1990s, these groups largely confined their drug trafficking to the Eastern Hemisphere, with a primary emphasis centering on the goal of making money while destroying infidels with opium, heroin, and related products. In more recent years, however, these groups have expanded their theater of operations to establish themselves in Latin America as well.

The notorious Colombian Cali drug cartel, for example, has long maintained mutually beneficial relationships with the pseudo-Marxist FARC and ELN (in Spanish, initials for the Revolutionary Armed Forces of Colombia and the National Liberation Army, respectively) guerillas. In exchange for the drug lords' money, the guerillas would murder uncooperative judges, excessively stubborn prosecutors, indiscrete journalists, uppity peasants, and other inconvenient people.

However, if it's terrorists you want, why settle for a bunch of amateurish Che Guevara wannabes when you can have Hezbollah? And if you are an Islamofascist group seeking to create a Western Hemispheric staging ground and recruitment center for assaults upon America, why not accept the invitation to do so, all expenses paid?

Liberal Western commentators have sometimes described the Islamist terrorists as being "puritanical" in outlook. Nothing could be further from the truth. In fact, as we shall discuss further in our final chapter, the moral basis of these groups is complete denial of conscience and the open embrace of pure depravity. Thus, for the drug lords, they are a perfect match. And the marriage has been consummated.

For example, the triborder region of Paraguay abutting Brazil and Argentina has become a major center of Islamist activity in South America. Tens of thousands of Mideastern Muslims have illegally migrated to the area, where more than a hundred hidden airstrips and dozens of terrorist training camps have been set up. According to the CIA, al Qaeda is highly active in the area, conducting arms-for-drugs deals with the FARC, the ELN, and the Peruvian Shining Path terrorist group.[13] On September 11, 2001, Brazil's former national drug enforcement czar, Jude Mangiello, revealed intelligence confirming this, adding that the growing local al Qaeda cells were involved in not only cocaine and arms trafficking, in conjunction with Colombian, Bolivian, and Peruvian drug dealers, but uranium smuggling as well.[14]

Hezbollah is also very active in the triborder region.[15] A tiny tip of the iceberg of their operations came into view in September 2001, when Paraguayan police raided warehouses in Ciudad del Este owned by one Assad Muhammad Barakat, who the police described as "the Hezbollah military chief in the tri-border region." The warehouse was found to be filled with training videos for suicide bombings and other explosive techniques, and motivational speeches by Hezbollah leader Hassan Nasrallah. Barakat got away but was arrested in Brazil, after which Argentine police revealed evidence connecting him to the bombing of the Jewish Community Center in Buenos Aires in 1994.[16]

In addition, however, Hezbollah has established terrorist cells and recruiting operations in Colombia, Panama, Ecuador, and Venezuela. Under Hezbollah auspices, for example, a large Shiite community has been established in the Colombian city of Maicao, which borders Venezuela, and turned it into a major center for black market, money-laundering, illegal arms, and drug-trafficking activities.[17]

This is just the beginning. In preparation for major expansion, the group has set up weekend camps throughout the area to train children and teenagers in weapons use and indoctrinate them in Islamist ideology.[18] With the advent of Iranian ally Hugo Chávez as dictator of

Figure 9.3. Hezbollah Venezuela poster. While America sleeps, Islamists are using their alliance with Latin American drug lords to establish bases in the Western Hemisphere. For a copy of their proclamation of jihad against America (in Spanish), see "Hezboallah llama a la Yihad en América Latina," http://groups.msn.com/autonomiaislamicawayuu/convocatoriaalayihad.msnw (accessed April 7, 2007).

Venezuela, such Hezbollah colonization operations can only be expected to accelerate rapidly.

Americans need to take the issue of Islamofascist expansion into Latin America very seriously. The two-thousand-mile US border with Mexico is notoriously porous and open to illegal immigration. Up until now, however, the primary concern with illegal immigration has been limited to relatively benign issues such as the potential impact of cheap Mexican labor on the US job market. However, once the Islamists become established in South America, it will not be merely well-meaning job seekers crossing the border, but lethal terrorists fanatically committed to our destruction.

It is the drug lords who have provided the Islamofascists with their invitation into our hemisphere. How can they be shut down?

As a criminal enterprise, the narcotics trade suffers from two critical weaknesses: Its ultimate foundation is both land and labor intensive. A lot of people have to be willing to devote themselves to growing the required crops, and they have to be willing to dedicate a great deal of land to that purpose.

Now the Islamist poppy farmers of rural Afghanistan may be proud that the poison they grow will be used to destroy infidels, but Latin American peasants growing crops for the drug traffic know that what they are doing is wrong.

So the question is: Why do they do it? "For the money" is the obvious answer, but it's a bit too quick. Working for a brutal criminal enterprise is personally degrading and quite risky, and most people would prefer not to have to do so. They do it because they *have* to—because it is the *only* way they can see to make a living.

The problem in rural Latin America, as throughout most of the third world, is the refusal of the advanced-sector nations to allow imports of their agricultural produce. It is this protectionist policy that, by denying many peasants a legal cash crop, forces them into the service of the narcotics trade.

The fact that this is the case has been shown by a modest US government initiative pursued under the Andean Trade Promotion Act

(ATPA, later renamed the Andean Trade Promotion and Drug Eradication Act, or ATPDEA).[19] Targeted toward Colombia, Ecuador, Peru, and Bolivia, this act gives duty-free treatment to certain agricultural imports as an incentive to encourage farmers to switch from coca to other crops. Unfortunately, the ATPA explicitly excludes sugar and most other major agricultural products from its purview, but the US Congress was able to muster the courage to rebuff the protectionist demands of the mighty American cut-flower industry and liberalize trade in at least this sector. As a result, the export of cut flowers from Colombia to the United States became highly profitable. This created an opportunity for many Colombian peasants to enter the legitimate economy, and hundreds of thousands of them were delighted to seize it.

Before the ATPA was enacted, Colombia was essentially kept out of the US flower market by high tariffs. After the ATPA passed, Colombia became America's main supplier of flowers. The industry has grown to the point where it is a significant economic sector and has been instrumental in helping the Colombia government regain control of several narcoterrorist regions.[20]

The ATPA is also credited with causing the elimination of some tens of thousands of hectares of coca farms in Peru and Ecuador as well. Evidently, despite the high street price of cocaine, the peasants who grow coca don't see much of the money, or at least not enough to make them choose it when there is a legal alternative available.

However, if we want to create an advanced-sector agricultural demand sufficient to replace the narcotics harvest, cut flowers are not going to do the job. A much larger market is required. This can be supplied by the alcohol economy.

This year, $600 billion will be shelled out by Americans, Europeans, and Japanese as tribute to OPEC, helping it fund terrorism. That same money, spent on alcohol crops, could convince a lot of narcotics farmers to grow something else.

By itself, a vigorous crop substitution initiative based on the alcohol economy could severely cut the narcotics harvest—but, admittedly, a significant fraction would remain. As most of the land was taken out of narcotics production, the price for the remaining product would go up, so some people would undoubtedly stay in the business out of pure greed. But under such conditions, where narcotics cultivation has become a minority enterprise rather than a necessary regional staple, crop eradication programs can become effective.

Every plant species has diseases or parasites specific to it that can be quite effective in its destruction. For example, in the 1940s and 1950s American elm trees were hit by an outbreak of the Dutch elm disease, and despite desperate defensive efforts backed by the full resources of the US government, they were virtually wiped out. Since the 1970s researchers have identified similarly effective fungi or other pathogens variously specific for marijuana, cocaine, and opium poppies.[21] The use of such mycoherbicides to destroy coca was signed into law in the United States on July 13, 2000.[22] They could easily be used to wreak havoc on narcotics crops worldwide, but have not been because of concern expressed by grower countries over massive socioeconomic impact to the affected regions.[23] However, once the alcohol economy has made substantial worldwide crop substitution possible, there would be no further reason to hold back. Species that currently provide the foundation of the global narcotics trade could be relegated to the status of curiosities preserved from extinction in greenhouses for the entertainment of professional botanists.

That's how we defeat narcoterrorism.

CHAPTER 10

DEFEATING GLOBAL WARMING

We have to offer up scary scenarios, make simplified, dramatic statements, and make little mention of any doubts we may have. Each of us has to decide what the right balance is between being effective and being honest.

—Dr. Stephen Schneider, leading proponent of global warming theory, *Discover* magazine, October 1989

No matter if the science is all phony, there are collateral environmental benefits. . . . Climate change [provides] the greatest chance to bring about justice and equality in the world.

—Christine Stewart, Minister of the Environment of Canada December 1998

The Book of Revelation [says] God will destroy those who destroy his creation. . . . Noah was commanded to preserve biodiversity. . . . Politics falls short of the minimum necessary to really address this crisis. . . . If you believe, if you accept the reality that we may have less than 10 years before we cross a point of no return—if you believe that, this is a time for action.

—Former vice president Al Gore New York City town meeting, May 23, 2006

Venus is a planet of steaming jungles, or so it was thought until the mid-twentieth century. Adventure writers like Edgar Rice Burroughs populated its tangled swamps with reptilian monsters, savage races, and beautiful princesses waiting to be rescued by earthmen brave enough to voyage there and try their luck. The scientific literature was not quite so imaginative, but as late as the early 1960s, many leading astronomers still did not rule out such wonderful possibilities.

The picture, after all, was not completely implausible. Venus is closer to the sun than Earth is, and so receives twice the solar radiance. But according to well-known scientific laws, the temperature of a body being heated by light rays increases only as the fourth root of the illumination power. In consequence, a planet getting twice the sunlight of Earth should have its absolute temperature increased by only 19 percent compared to ours. Working the numbers, this would suggest that typical temperatures on Venus should be about 64°C, or 147°F.[1] That's hot, but not impossibly hot for life. Furthermore, that's just the average planetary temperature. Surely some places, such as the poles, could be much cooler. The Venusian Arctic might be just right.

Alas, when the American spacecraft *Mariner II* reached Venus in 1964, it took measurements from orbit that showed the planet's temperature to be not 64°C, but at least 330°C (600°F)! This was incredible, but if anyone doubted it, in 1967 the Soviet probe *Venera II* more than confirmed the finding by entering the atmosphere and parachuting down to the planet's surface. It didn't last long, but before it died it took direct temperature readings showing the surface of the planet to be baking away at 480°C (900°F). That's hot enough to melt lead. As the noted astronomer Carl Sagan put it at the time, "Venus is hell."

The discovery was astonishing, but it didn't take scientists long to explain it. The planet named for the goddess of love is surrounded by a very thick carbon dioxide atmosphere. One property of carbon dioxide molecules is that they strongly absorb infrared radiation, which is the type of radiation given off by hot objects. A thick carbon dioxide atmosphere would thus act as a kind of greenhouse roof over the planet, letting sunlight in to warm the surface, but then trapping the heat below before it could escape back into space.

Thus was born, in popular parlance, the "greenhouse effect." A number of scientists had actually speculated on the possibility, even before *Mariner II*. Their arguments had been disputed, however, and certainly no one had suspected that the effect would be so strong. But seeing is believing, and after *Mariner II* there could be no doubt. The greenhouse effect was real.

Fortunately, the discovery did not start a panic. The atmosphere of Venus contains three hundred thousand times as much carbon dioxide as Earth's, so the scientists knew that, short of the sun exploding, no disaster comparable to the Venusian greenhouse was going to happen here. It was, of course, well known in the 1960s that our industrial society was producing increasing amounts of carbon dioxide emissions. However, Earth at that time was actually undergoing a cooling trend, so the climatological doomsayers of the day were predicting a new ice age.[2] Global warming was not an issue.

In the 1970s, however, worldwide temperature measurements started to increase. This trend continued, so that by the next decade many leading climatologists were warning of a threat to society posed by a man-made greenhouse effect.[3]

Since the 1980s the view that global warming is a real danger has become increasingly accepted in many policy-making circles, leading to the signing of an international treaty in Kyoto in 1997 that would impose drastic antigreenhouse countermeasures as necessary policies upon all its signatories.

Because such efforts to counter the greenhouse effect necessarily would require massive government intervention into the economy, not only they, but the global warming hypothesis itself, have become highly controversial. With as much predictability as the lineup of congressional votes during the Clinton impeachment trial, individuals and factions generally partial to greater state control of economic life have found the evidence for global warming to be quite compelling, while those of more libertarian persuasions have dismissed it as little more than a hoax.[4] For the Malthusians, global warming has proven a godsend, actual proof that humanity's aspirations must be decisively constrained lest we destroy Nature. Seizing this line as his bandwagon, former vice president Al Gore has gone so far as to make dramatic proclamations that "we have just ten years left to stop global warming" and has produced an agitational movie and matching book whose contents leave no doubt as to the identity of the Savior on Horseback who must be empowered if the world is to be rescued from our folly.[5]

The stakes in this debate are quite high. If global warming is a reality and we do nothing to avert it, it could wreak severe havoc. For example, a sharp temperature increase could melt the glaciers of Greenland and Antarctica, resulting in a sea level rise of as much as 70 m (230 ft). This would flood many low-lying coastal areas, with massive economic damage resulting. On the other hand, government regulations along the lines of the Kyoto agreement to stop greenhouse warming by forcing the reduction of carbon dioxide emissions could stifle worldwide economic growth, thereby causing millions of deaths through the perpetuation of global poverty.

While politicians may well be entitled to believe whatever they please on a variety of issues, global warming is either real or it isn't. It's not a matter of opinion or what is convenient to whose ideology. With the potential consequences of a mistaken response being so serious, we need to get it right. So let's put aside factional spin and have a look at the facts.

IS GLOBAL WARMING REAL?

It's hard to know the "global temperature," since the actual temperature is always different from place to place, and the average temperature derived from a large number of stations can be severely skewed by the choice of location of the measuring stations. Nevertheless, serious efforts to assign an average global temperature through systematic worldwide thermometer readings have been made since about 1858. The data from this century-and-a-half-long global research program initiated by the British Empire at its Victorian height and continued down to the present day is summarized in figure 10.1.[6] This is the best historical climate change data set that there is, and it forms the basis of all the official analysis and policy documents issued to the UN Intergovernmental Panel on Climate Change (IPCC) and similar bodies. It can be seen that there actually has been an increase in global temperature of about 0.8°C (1.4°F) over the period in question.

So, as measured empirically since the mid-nineteenth century, global warming is certainly real, but it's not that fast. A temperature increase of 0.8°C over 145 years works out to a rate of just 0.55°C (1°F) per century. That's not particularly alarming. However, it may be noted that in the past twenty years the pace seems to have picked up

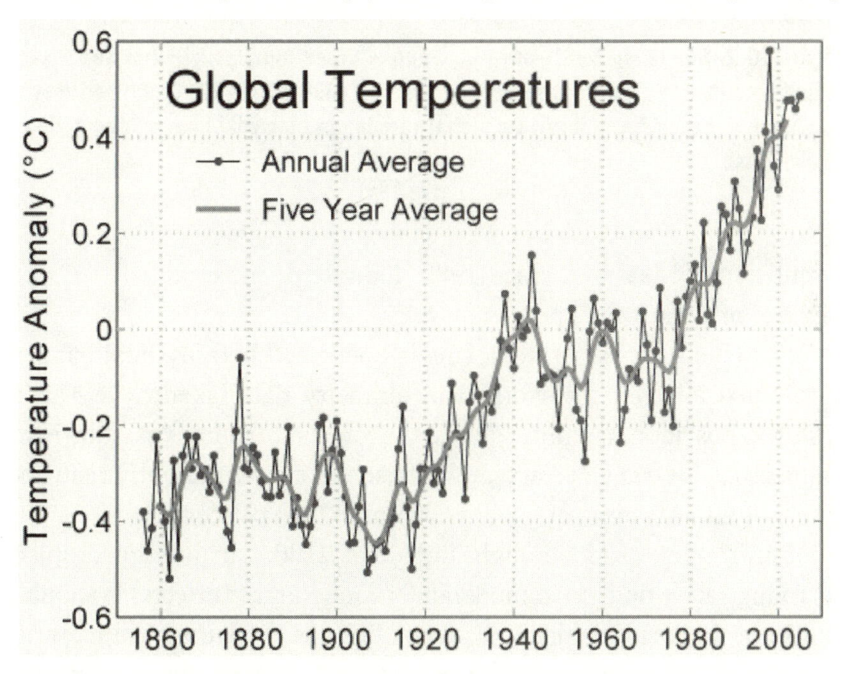

Figure 10.1. Global average temperatures. ***Source*****: Data from the Hadley Centre for Climate Prediction and Research of the UK Meteorological Office. Graph courtesy of Global Warming Art, http://www.globalwarmingart.com/wiki/Image :Instrumental_Temperature_Record_png.**

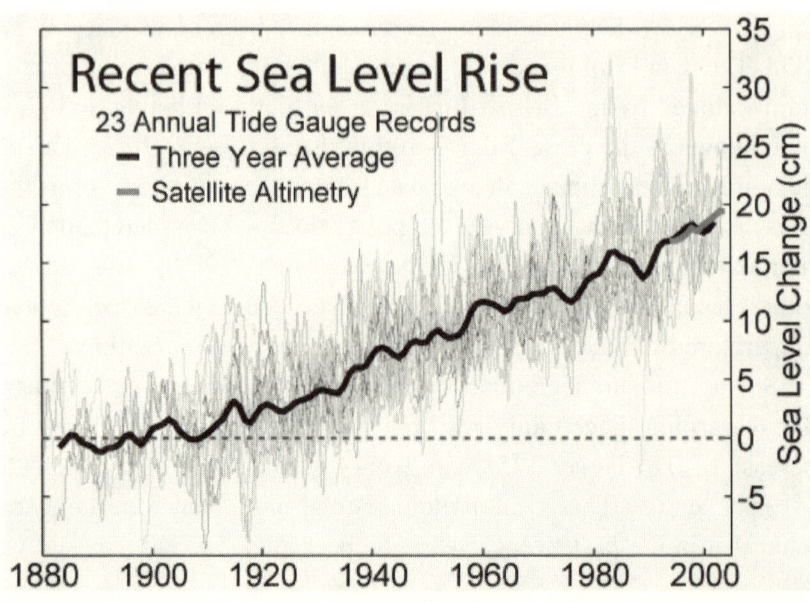

Figure 10.2. Sea level, 1880–present. ***Source*****: Bruce Douglas, "Global Sea Rise: A Redetermination,"** ***Surveys in Geophysics*** **18 (1994): 279–92. Graph courtesy of Global Warming Art, http://www.globalwarmingart.com/wiki/Image:Recent_Sea _Level_Rise_png.**

significantly. Measured since 1980, the rate of increase is 0.2°C per decade, or 2°C (3.6°F) per century.

Sea levels are rising, too. This is confirmed both by tidal gauges positioned all over the world and altimetry data taken by NASA's TOPEX/Poseidon satellite. The data from both sources is shown in figure 10.2.[7] It can be seen that world sea levels are currently rising at a rate of about 3 mm per year, or 30 cm (12 in) per century.

So, if continued through the year 2100, such a quadrupled warming rate would have moderate but measurable effects. It should, however, be evident that these measurements give no support to alarmist assertions that we have only ten years to stop global warming. In ten years, at current rates, global temperatures will have increased only a further 0.2°C, and sea levels will rise 3 cm (1.2 in), neither of which is remotely enough to matter.

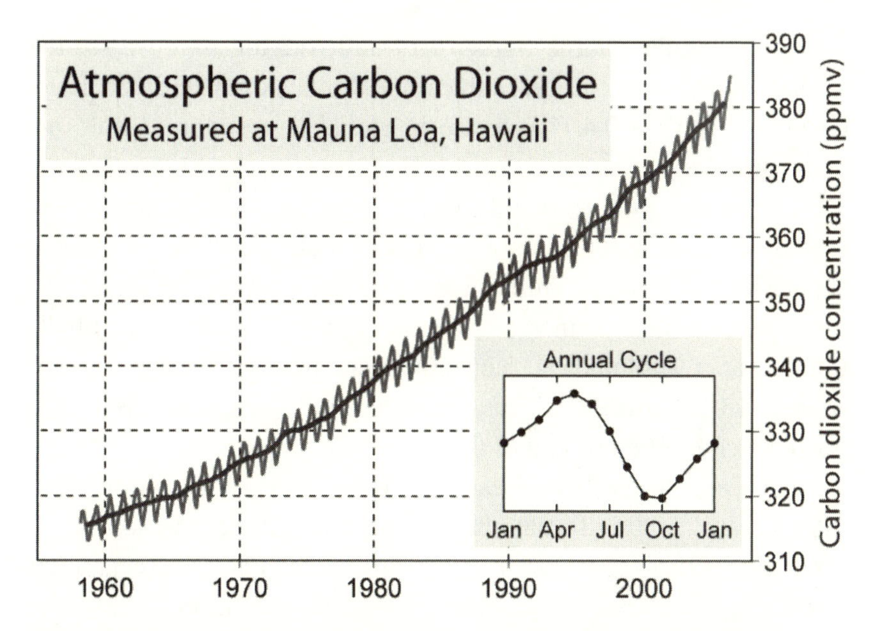

Figure 10.3. Record of the carbon dioxide in the atmosphere, 1958–2002. ***Source*: Data from Mauna Loa Observatory. Graph courtesy of Global Warming Art, http://www.globalwarmingart.com/wiki/Image:Mauna_Loa_Carbon_Dioxide_png.**

The next question is whether this warming trend is actually being caused by increased carbon dioxide in the atmosphere. That's hard to answer definitely, because there are many other factors that affect climate. However, as shown in figure 10.3,[8] there is no doubt that global carbon dioxide atmospheric levels are actually increasing, and in fact have risen about 20 percent since 1958.

The record shown in figure 10.3 looks like saw teeth because the amount of carbon dioxide in the atmosphere changes over the course of a year. Green plants absorb carbon dioxide during the spring and summer when they are actively growing. Because there is more land area in the Northern Hemisphere than the Southern Hemisphere, carbon dioxide concentration increases and decreases in accord with the northern seasonal cycle. Thus these concentrations peak each year in March, just before the growing season starts in the north. We shall return to this important point later.

So, taking the data in figures 10.1 and 10.3 together, we see that we have a 20 percent increase in carbon dioxide over the same period that we observe a 0.4°C increase in global temperature. Could one have caused the other?

The answer is quite possibly. According to the mathematical models used by climatologists, a doubling of carbon dioxide in the atmosphere should lead to a temperature increase on the order of 1.5 to 5°C.[9] If that's the case, than an increase of a factor of 1.2 (which is roughly the fourth root of 2) should cause a temperature increase of about a quarter of this, or 0.38 to 1.25°C. Thus, the observed 0.4°C increase is a pretty good match for the low end of the theoretical predictions.

If we extend the rate of carbon dioxide increase shown in figure 10.3 forward in time for another century, we get a prediction of 575 parts per million (ppm) for the year 2100, which is 50 percent above today's level and roughly double the preindustrial level of 280 ppm. Using a midrange value of a temperature increase of 3.2°C per doubling of atmospheric carbon dioxide, this would imply a temperature increase of 1.8°C over the course of the next century, in pretty good agreement with our previous estimate of 2°C.

In going from 315 ppm carbon dioxide in the atmosphere in 1958 to 375 ppm today, we increased the total mass of carbon (as opposed to carbon dioxide) in the atmosphere from 630 billion tonnes (or gigatons, Gt) to 750 Gt. Since coal is nearly all carbon and natural gas and oil are 75 to 87 percent carbon, producing such an increase artificially would have required us to burn a minimum of 140 Gt of fossil fuels between 1958 and today. In fact, the actual amount of fossil fuels used worldwide over that period was around 250 Gt.[10] Thus it would appear that we may well have been responsible for the observed increase, and would have caused a lot more except for the ability of the biosphere to take in much of our carbon dioxide emissions and turn it into plant material.

So, as well as such things can be proven, it's a fact: Humans have changed and are changing the climate of Earth. But before you run out and enlist with Al Gore, I ask that you stop and take a look at the data presented in figures 10.4, 10.5, and 10.6. Here we see results gathered

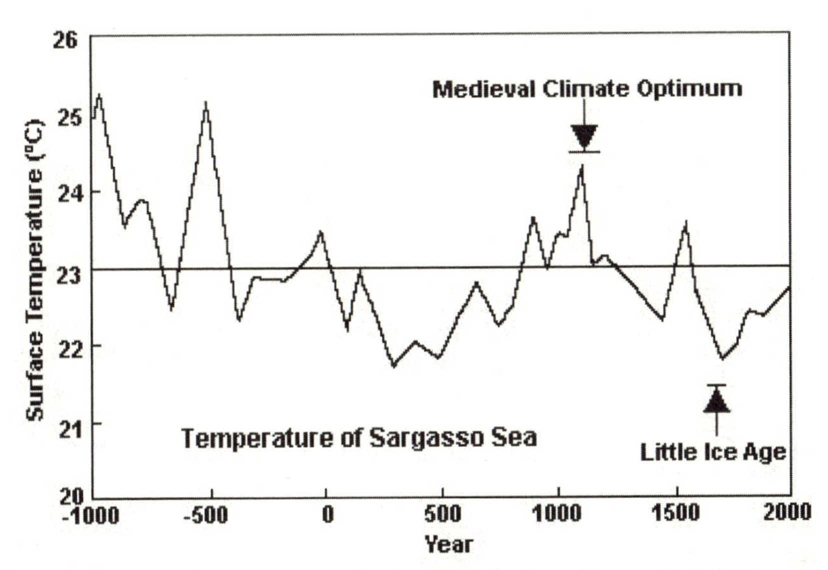

Figure 10.4. Surface temperatures in the Sargasso Sea. *Source*: A. B. Robinson, S. L. Baliunas, W. Soon, and Z. W. Robinson, "Environmental Effects of Increased Atmospheric Carbon Dioxide," *Climate Research* 13 (1999): 149–64. Reprinted with permission.

by various teams of researchers for much longer periods of history than that measured by the thermometers deployed by the UK meteorological office since 1858.

Let's start with figure 10.4.[11] This shows the temperature of the Sargasso Sea as determined by isotope ratios of marine organism remains.[12] These fossils, preserved in sediments on the seafloor, provide us with a temperature record that covers not 150 years, but 3,000 years. This data, which is similar to other studies of various world locations, shows a period of higher temperatures 1,000 years ago ("Medieval Climate Optimum") and lower temperatures 300 years ago ("Little Ice Age"). The horizontal line is the average temperature for the 3,000-year period.

Global warming may be a fact, but it is also a fact that a thousand years ago the world was significantly warmer than it is today. A thousand years ago, the snow line in the Rockies was a thousand feet higher than it is now, and Canadian forests flourished tens of kilometers farther north. A thousand years ago, oats and barley were grown in Iceland, wheat in Norway, hay in Greenland, and as far north as York, the vineyards of England produced fine wines.[13] These warm temperatures were no disaster. On the contrary, persisting through the twelfth century, they are believed by historians to have contributed materially to the significant growth of population and prosperity in Europe during the High Middle Ages (roughly the years 1000 to 1300).

In figure 10.5, we show composite data for the past eighteen thousand years, according to data assembled by the UN Intergovernmental Panel on Climate Change.[14] Here we see a temperature rise of about 6°C from the end of the last ice age to the period of the dawn of civilization, about six thousand years ago (i.e., 4000 BCE).

In figure 10.6, the history of Earth's climate since the end of the age of dinosaurs is shown, based on data derived from examining oxygen isotopic ratios in seafloor material.[15] It can be seen that the overall trend has been a cooling one, with temperatures dropping 12°C (21°F) over the past 50 million years.

If we consider figures 10.4, 10.5, and 10.6 together, two remarkable generalizations stand out. First, we see that the natural variations

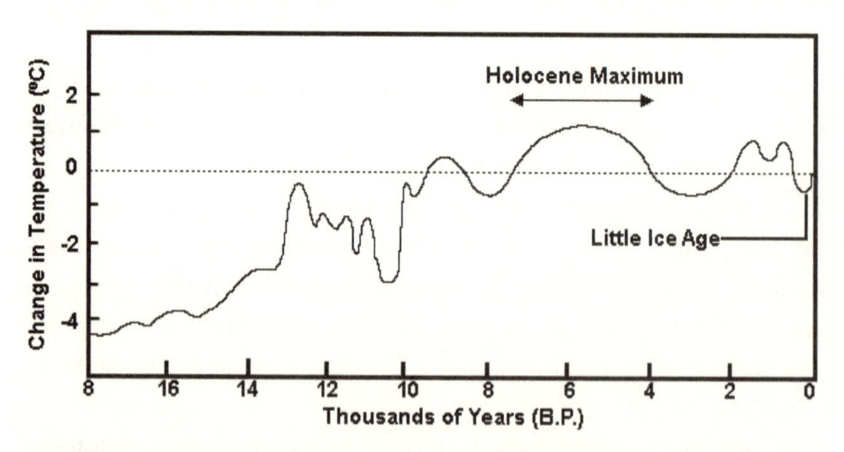

Figure 10.5. Average global temperature since the last ice age. ***Source*****: R. S. Bradley and J. A. Eddy,** ***EarthQuest*****, vol. 1, 1991. Based on J. T. Houghton et al.,** ***Climate Change: The IPCC Assessment*** **(Cambridge: Cambridge University Press, 1990), http://gcrio.org/CONSEQUENCES/winter96/article1-fig1.html.**

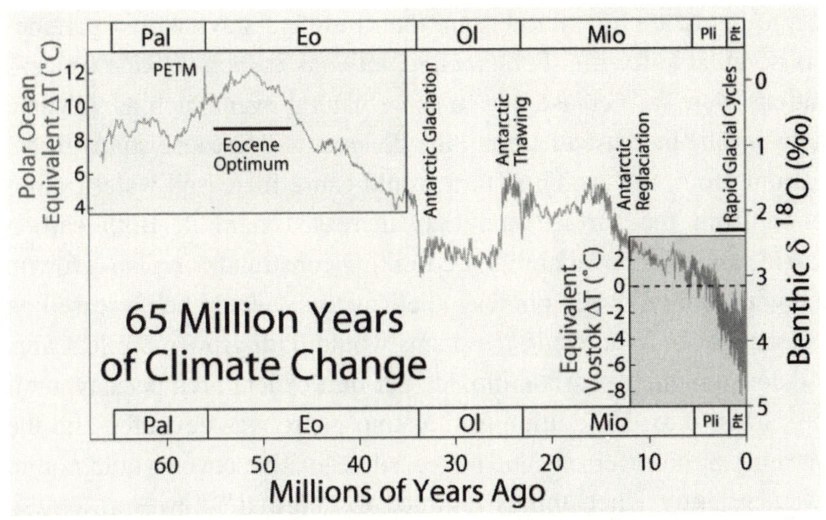

Figure 10.6. Deep ocean temperatures over the past 65 million years. *Source*: Data from James Zachos, Mark Pagani, Lisa Sloan, Ellen Thomas, and Katharina Billups, "Trends, Rhythms, and Aberrations in Global Climate 65 Ma to Present," *Science* 292, no. 5517 (2001): 686–93. Graph courtesy of Global Warming Art, http://www.globalwarmingart.com/wiki/Image:65_Myr_Climate_Change_Rev_png.

in Earth's temperature exceed man-made global warming to date by a long shot.

Second, we see that despite these significant natural temperature wanderings, in the largest scheme of things, Earth's temperature does not change very much. This can only imply that, considered as a system, Earth's climate is *stable*.

Stable systems are those that, when perturbed by outside influences, seek to return to their original condition. Unstable systems, in contrast, respond to perturbations by increasing the disturbance. By way of analogue, think of a ball in a crater as representing a stable system, but one on top of a hill as an unstable one. Kick the stable ball in any direction and it will roll back to its original position. Kick the unstable ball and it will roll off the hill.

Earth's climate may get kicked out of place from time to time, but it always rolls back. This is a fundamental feature of the biosphere; it acts to regulate climate to ensure that it always stays within parameters acceptable for life. For example, let's say carbon dioxide concentrations were to increase due to some natural event such as volcanic activity. The increased carbon dioxide level would cause some global warming to occur, which in turn would cause increased water evaporation from the surface and thus increased rainfall. Both carbon dioxide and water availability are limiting constraints on plant growth, and with more of each, photosynthetic activity would be increased on a global scale. As a result, the plants would both grow more leaf area and consume more carbon dioxide per unit of leaf area per day, until the carbon dioxide addition to the atmosphere was devoured and the warming trend reversed. The increased vegetative cover would not go away instantly when things returned to "normal," however, and so carbon dioxide levels would generally be pushed below their original level. This would go on until the resulting global cooling decreased vegetative activity and carbon dioxide levels, and thus temperature, had a chance to rebound.

Thus the system can oscillate, like the ball rolling around the basin of a crater, but it always stays within fairly well-confined parameters. There is some change, however, over eons of geologic time. This is because the biosphere is evolving. For example, looking at figure 10.6 we see that the basic trend over the past 65 million years is one involving significant global cooling since the time of the Paleocene-Eocene Thermal Maximum (PETM), whose ~2,000 ppm carbon dioxide atmosphere supported a global climate that was 12°C warmer than our own.[16] During the same period, new types of plants appeared that were capable of using carbon dioxide more efficiently. The most important of these were the grasses, which, originating at the end of the Paleocene (about 58 million years ago), drove temperatures sharply down from that point on (see figure 10.6). Things then stabilized in the Oligocene, until the mid-Miocene, about 15 million years ago, when new types of grasses appeared. These more advanced grasses, dubbed C_4 as compared to the previous C_3 varieties, were much more efficient at carbon dioxide utilization, and, as they replaced their predecessors, they drove atmospheric carbon dioxide levels below 300 ppm for the first time in Earth's history.[17] In the process, they sent the planet's climate plunging into a glacial age that has continued to the present day.

The fact that Earth's atmospheric carbon dioxide concentration was about 280 ppm shortly before the industrial revolution implies that this value is close to the equilibrium level for the modern biosphere. As human industrial activity raises the carbon dioxide levels significantly higher, we will increase plant productivity, causing the biosphere to push back like a spring, with the force of its push becoming ever stronger the further the system is displaced from equilibrium.

So, if global warming is really occurring at a rate too slow to make it of any immediate urgency, if the continuation of global warming at its current rate for another century would just return us to a climate that humans found quite attractive in the past, and if the biosphere possesses restorative powers capable of resisting excessive deviation from climatic norms, should we still care about the greenhouse effect?

Yes, we should.

WHY WE NEED TO ADDRESS GREENHOUSING

The reason why we need to think about the effects of greenhouse gases is *not* because they represent any near-term threat of a global warming catastrophe. As we have seen, such alarmist incantations are simply untrue. We certainly don't need a would-be Savior on Horseback hyping such nonsense to rescue us from The Wrath to Come. There is no impending doom. Put simply, humanity's current rates of carbon dioxide emissions are benign. If continued at present levels for another century, they would warm the world at a modest and ever-decreasing rate, increasing rainfall and plant productivity until the climate stabilized in the neighborhood of its high medieval optimum.

However, that said, we cannot continue with current levels of carbon dioxide emissions—because they are much too low.

We live in a very unjust world. Even the United States, with a per capita GDP of $42,000 per year, still is afflicted with significant poverty. Yet the average worldwide per capita GDP is $8,800, and billions of people live in countries where it is less than $1,000. If we are to attain a global society where every person, or even most people, have a decent standard of living, we are going to have to multiply worldwide economic activity many times over. That will require a parallel increase in energy utilization.

Many American politicians have objected to the Kyoto treaty because, while it demanded that the carbon dioxide emissions of advanced-sector nations be reduced to and then frozen at or below 1990 levels, it put no restrictions on the emissions of the developing countries. Actually, that was the only good feature of the treaty. Had it forced the third world to accept the same terms it would have been a genocide pact, destroying the lives of billions of people by immobilizing their nations in perpetual poverty. Not even African kleptocrats were willing to accept that. Accordingly, an alternative was proposed in which third world nations would be given carbon dioxide emissions limits somewhat above their current levels, and, to the extent that they produced less than these amounts, they could sell "carbon credits" for such shortfall to advanced-sector nations whose fuel use exceeded their Kyoto quotas. Had this concept gone forward, it would have marked a new low in human depravity, as it effectively would have institutionalized an international system in which rich countries paid huge sums of money to third world dictators in return for keeping their countries poor. Fortunately, however, the kleptocrats, while liking the idea in principle, could not agree on their own quotas, so the idea was stillborn.

As noted, the United States is currently far from an economic utopia. Yet let us take it as our baseline for an acceptable worldwide living standard to be achieved by the end of the twenty-first century. Such a goal is not fantastical. In fact, the world, which has always been much poorer than the United States, has today achieved a per capita GDP comparable to that prevalent in America in 1900 (US 1900 GDP per capita is estimated at about $5,000 in today's dollars).[18] Further, let us assume, in accord with many projections, that the population of the world will also double over the next hundred years. This assumption may be conservative: During the twentieth century the world population actually quadrupled. But that's what many demographers predict, so we'll go with it.[19]

If we combine the fivefold increase required to raise global per capital GDP with a doubling of population, we arrive at the necessity of increasing worldwide GDP tenfold over the coming century. Again, such a goal is not fantastical. It is the equivalent of an average worldwide rate of economic growth of just 2.4 percent year (it averaged 3.6 percent over the twentieth century and 3.2 percent between 2000 and 2005). If patterns of energy consumption remain the same, with worldwide fuel use per unit of GDP matching the current worldwide average, this would multiply global carbon dioxide emissions tenfold as well.

We are currently putting around 9 Gt per year of carbon into the atmosphere. As we can see in figure 10.3, this is causing a net rise of about 1.6 ppm carbon dioxide in the atmosphere per year, or about 0.4 percent. At this rate it would take 230 years to achieve a carbon dioxide doubling that would result in a 3°C temperature rise. This is not a big affair, especially when you consider the likelihood of expanding vegetation induced by the carbon dioxide increase slowing and then stopping the rise well before matters get that far.

However, the effect of dumping 90 Gt of carbon per year into the atmosphere would be quite another thing altogether. It would, in fact, be significantly more than ten times as great as our current 9 Gt per year level, because, as can be seen from looking at figure 10.3, currently plants can take out most of what we put in. That is, each year the carbon dioxide level actually rises about 8 ppm, but plants take out around 6.4 ppm, or roughly 75 percent of the gross increase that would otherwise occur. Thus we only see a yearly net increase of about 1.6 ppm. But if instead we were putting out 90 Gt of carbon per year, the gross rise would be more like 80 ppm.

It is a fact that increasing carbon dioxide levels will increase plant productivity. At preindustrial levels of 280 ppm, most plants are operating in a carbon dioxide–deprived condition, so adding more to the atmosphere removes a limiting factor to their growth. Satellite observations have revealed that concurrent with the 19 percent increase in atmospheric carbon dioxide concentration since the 1950s, the rate of plant growth in the continental United States has increased by 14 percent.[20] Studies done at Oak Ridge National Lab on forest trees have shown that increasing the carbon dioxide level 50 percent higher, to the 550 ppm level projected to prevail at the end of the twenty-first century, will likely increase photosynthetic productivity a further 24 percent.[21] By raising the temperature and thus vapor pressure of the world's oceans, and also promoting the expansion of leaf area and thus plant transpiration, global warming will increase worldwide rainfall. This will accelerate crop growth as well.

Yet in the absence of well-directed human assistance, even a significantly enhanced biosphere could have a rough time compensating for an artificial carbon release of 90 Gt per year into the atmosphere. Under such circumstances, net carbon dioxide atmospheric increases on the order of scores of ppm per year would not be unthinkable. At such rates, it could take only a few decades to reach Eocene carbon dioxide atmospheric concentrations of 2,000 ppm, with really significant temperature rises, flooding, or other serious consequences ensuing.

Timing is everything. In life, we face many problems that all need to be solved, but on different time scales. Speaking in round terms, we need to breathe every minute, drink water every day, eat food every week, pay our bills every month, find work every year, and set aside money for our retirement over decades. All of these problems are important, but some have priority over others. If you are a scuba diver deep underwater and you are running out of air, you would do well to focus on the problem at hand rather than dwelling on your mortgage payments or retirement fund.

Similarly, in addressing the energy problem, we face three major issues. In order of urgency, these are strategic, economic, and environmental. The strategic issue is the most urgent because we need to move fast to defund and disarm people who are trying to kill us. The economic issue is second, because it concerns obtaining the means to sustain and develop our society. The environmental issue is third, because only after the means of existence have been secured can we meaningfully address the question of the quality of that existence.

Within the environmental domain, we also have to prioritize. Global warming will absolutely have to be addressed in due course. However, of much greater immediate urgency is the massive expansion of conventional pollution due to the growth of coal and oil utilization by the developing nations. India and China, in particular, are already encountering severe public health consequences from the vast brown clouds of nitrous, sulfurous, and particulate pollution created by their fossil-fueled industrialization. Unless we can define a long-term energy policy that mitigates such effects, the kind of general development needed to lift the crushing burden of third world poverty will raise lung cancer, asthma, and other respiratory afflictions to a global curse.

So, the strategic, economic, and environmental dimensions of the energy problem all need to be solved, but each on its own time scale. If using some coal to make methanol cheaper than oil is necessary to break OPEC, we can do that, even though coal-derived methanol does nothing to fight global warming. The key immediate task is to drive the transition to an alcohol economy. Once we do that, we will be in a position to put world development on a technological foundation that will allow progress to go forward in a way that is fully compatible with all of our hopes for human and environmental happiness.

Turning, therefore, to a longer-term perspective, let's see how this can be done.

CHAPTER 11
EMPOWERING THE FUTURE

There are no ends, limits or walls that can bar us or ban us from the infinite multitude of things.

—Giordano Bruno, *On the Infinite Universe and Worlds*, 1584

We believe that free labor, that free thought, have enslaved the forces of nature, and made them work for man. We make the old attraction of gravitation work for us; we make the lightning do our errands; we make steam hammer and fashion what we need. . . . The wand of progress touches the auction-block, the slave-pen, the whipping post, and we see homes and firesides, and schoolhouses and books. . . . Swear that you will never vote for any enemy of human progress.

—Colonel Robert G. Ingersoll, Vision of War speech, 1876

How can we empower an unlimited future? To answer this question, we must expand our view of energy.

Thus far we have primarily devoted ourselves to the issue of producing liquid fuels for transport, because this is the key to both the current strategic situation and the potential to launch a new age of true world development. As we have seen, such fuels can be produced without a net addition to global warming by switching from fossil petroleum to alcohols derived from biomass. As the century progresses, the development of genetically engineered plants and cellulosic ethanol technology will expand our capability in this area much further. However, transportation represents only about 40 percent of all energy use. Another 40 percent is used to produce electric power, while the remaining 20 percent goes to various residential and industrial applications. If we are to adequately address the issue of preserving the environment during an age of rapid global economic expansion, we need to consider how we will meet these needs as well.

We have the technological capacity today to produce electricity without any air pollution or greenhouse emissions by making use of nuclear fission, geothermal, hydroelectric, solar, or wind technology. Provided we undertake the necessary research, in the future we will also have the capacity to generate virtually unlimited power by making use of thermonuclear fusion. However, the current reality is that roughly 66 percent of the world's electricity comes from burning fossil fuels; 16 percent each from nuclear and hydroelectric; 1 percent from burning wood, other biomass, or waste; 0.65 percent from geothermal; 0.31 percent from wind; and only 0.04 percent from solar. How do we get from where we are today to our goal of a truly prosperous world whose further development can proceed as far as desired without fear of environmental catastrophe?

The key to the transition is how we handle coal. Of the 66 percent of the world's electricity that is derived from fossil fuel combustion, 40 percent comes from coal, 19 percent from natural gas, and 7 percent from oil (see table 11.1). Furthermore, in addition to supplying most (61 percent) of the fossil fuel–generated electric power, and most of the world's industrial process heat, coal produces much more pollution as well as 50 to 100 percent more carbon dioxide per unit of power than do oil or natural gas, respectively. Of all the greenhouse emissions worldwide produced by electric power generation, three-quarters result from burning coal.

Because of coal's dominant role in carbon dioxide emissions, many global warming hawks have called for punitive measures, such

TABLE 11.1. WORLD MARKET FRACTION AND TYPICAL COST OF ELECTRIC POWER SOURCES[1]

Source	Market Fraction (percent)	Typical Cost (cents/kW-hour)
Coal	40	4
Natural Gas	19	5
Oil	7	6
Nuclear	16	5
Hydroelectric	16	2
Geothermal	0.65	6
Biomass	1	6
Wind	0.31	7
Solar	0.04	50

as costly "carbon taxes," to drive it out of the electric power market.[2] This, however, is the wrong approach. Coal is widely used because it is cheap—roughly one-tenth the cost per unit mass as oil or natural gas. The fact that it is cheap is good. It lowers the cost of electricity and, thus, everything made by electricity. That is a major boon to humankind that should not be wantonly thrown away due to hysteria over global warming rates of 0.2°C per decade and sea level increases of 12 inches per century.

There is a real environmental problem with coal, however, and that concerns its role in fomenting conventional pollution. The burning of coal in unsophisticated facilities can (and does) produce vast quantities of soot and ash particulates, acidic oxides of nitrogen and sulfur, and highly toxic trace elements, including mercury, selenium, and arsenic. This stuff is poison to humans and nature alike. It causes lung cancer and asthma. It also causes acid rain. Particularly in Asia, brown clouds of pollutants from coal combustion are turning many major cities into noxious atmosphere disaster zones. According to H. Keith Florig, a leading researcher at Carnegie Mellon University, air pollution-induced respiratory ailments are now killing more than a million people every year in China alone.[3] Conditions in India and elsewhere in Asia are reported to be nearly as bad.

These problems are very serious, but they can be dealt with by using proper technology. Ash, soot, and other particulates can be removed from the exhaust of coal-fired generator stations by fabric filters and electrostatic precipitators. Desulfurization techniques can eliminate 99 percent of the sulfur dioxide from the exhaust, and, together with the particulate control systems, eliminate the toxic trace elements as well. Nitric oxides can be cut 90 percent or more using advanced combustion technologies and by catalytic conversion of the flue gas. As a result of the introduction of these cleanup systems into many coal-fired power plants in the United States since 1980, total coal-induced pollution emissions have been reduced radically, even though during the same period the amount of coal being used nationwide increased by 74 percent.[4]

Instead of a regressive "carbon tax" to destroy the coal-fired power industry, what is needed is simply regulation requiring all coal plants, steel mills, and other coal-using industrial facilities to make use of modern cleanup systems. Diplomatic and, if necessary, aid efforts should also be employed to get other countries to adopt the same standards as well (the United States is way ahead of most of the rest of the coal-burning nations in this area).

While residential coal users cannot be expected to install such advanced technology, most of the pollutants can be removed from coal in advance of consumer distribution simply by washing the coal powder with water before it is formed into the standard briquettes that constitute its most common retail form in China and other developing nations. In the longer term, residential pollution from burning coal (as well as wood, dung, and other crude sooty sources) can be eliminated entirely by turning the coal into methanol or DME at well-managed industrial facilities, and then distributing these clean-burning fuels for home cooking and heating use.

Provided such measures are done, the world can continue to reap the benefits of its vast coal resources without receiving their potentially noxious curse.

In due course, however, we will also have to deal with the greenhouse gas emissions from coal combustion. The way this can be done is by implementing a more advanced type of clean-coal technology known as the Integrated Gasification Combined Cycle (IGCC), which is already operational in a few pilot plants. In the IGCC system, the coal is first reformed by reaction with water to form carbon dioxide and hydrogen, which are separated. The hydrogen is then piped off to be burned in a turbine and generates very clean electric power (this is very different than the "hydrogen economy," in that the IGCC hydrogen does not have to be compressed, liquefied, stored, or transported), while the carbon dioxide is sent through a scrubber. Once cleaned, the nearly pure carbon dioxide can be compressed to liquid form (carbon dioxide is a liquid at room temperature and 700 psi) and then sequestered underground. This latter process does not need to be an economic burden on the system. In fact, it can produce a profit, because liquid carbon dioxide can be (and is) used to push oil out of the ground from partially depleted wells.

While varying from place to place, a typical oil well recovers only about 30 percent of the oil that is originally present underground. Using secondary recovery techniques such as pumping water underground to put pressure on the remaining petroleum, the yield can be increased to around 50 percent. But if that is all that is done, half of the original subsurface supply will still be untapped. This is a very large resource. Nationwide, the amount of petroleum remaining in such "spent" American oil wells is estimated to exceed the total reserves of Saudi Arabia. In order to get at this petroleum, in recent years American oil companies have developed tertiary recovery techniques in which carbon dioxide, obtained by tapping natural underground reservoirs, is pumped into wells, where it dissolves into the oil and both pressurizes it like soda pop and reduces its viscosity, thereby enabling yet another 20 percent of the well's petroleum to be forced to the surface. As a general rule, it takes around 5,000 to 10,000 standard cubic feet (270 to 540 kg) of carbon dioxide to recover one barrel of petroleum, and with other costs included, oil companies are willing to undertake such an operation if they can get the necessary carbon dioxide supply for $2 per thousand cubic feet (i.e., $10 to $20 per barrel) or less. Unfortunately, natural underground carbon dioxide pockets exist only in certain locations, so application of the technique has been limited to wells found within short distances of such reservoirs. However, by laying pipelines from IGCC plants to areas of spent oil fields (which can be found in abundance in Pennsylvania, Texas, and California, to name just a few areas close to cities with large demands for electric power), the carbon dioxide exhaust can be made available for such purposes and sequestered at a profit.

While not as high a priority for such application because they produce less carbon dioxide and far less pollution per unit of power than coal, other materials—including oil, natural gas, methanol, dimethyl ether, and even raw biomass—could all also be used as fuel in IGCC systems.

The total amount of carbon dioxide that can be profitably sequestered in this way is very large, but not infinite. However, the half century or so during which it can be employed will give us the time needed to transition to an electricity sector dominated by sources that produce no carbon dioxide emissions at all.

By far the cheapest source of zero-emission electricity is hydroelectric (see table 11.1), which is quite literally in a competitive class by itself. If you have an appropriate watershed and terrain configuration, once built, a hydroelectric dam is practically a free lunch. Unfortunately, however, the number of good hydroelectric sites is intrinsically limited, and many of the best have already been exploited. The total hydroelectric power potentially available worldwide is only about three times that already being generated. Global warming over the next century might improve this picture by 14 percent or so (since the vapor pressure of the oceans, and thus worldwide rainfall, will increase by 7 percent for every degree centigrade of temperature rise), but even so, the role of hydro in an effort to meet the needs of a world economy expanding tenfold must necessarily be modest. Practical application of geothermal heat to generate electricity is even more geographically and geologically circumscribed.

Thus we are left with the options of nuclear, wind, and solar power. While passionately beloved by many people for its apparent elegance, from an economic point of view solar electricity is a non-starter. It is true that for the past several decades the cost of photovoltaic power has fallen by about a factor of two per decade, so *if* this trend were to continue for three more decades, and *if* other alternatives were to stand still during that period, solar might eventually become competitive.[5] However, as photovoltaics are nearing maturity, such an assumption would be quite a logical stretch.

In contrast to solar energy, wind power is almost competitive today. With the help of a moderate government subsidy of 1.7 cents per kW-hour, commercial wind farms are growing in the United States. Across Europe, strong government support has enabled wind power to grow to 2.4 percent of the European Union's power generation capacity. The success of wind has been particularly noteworthy in Denmark. In that country, where high-wind locations are readily available near the most important population centers, wind now supplies 20 percent of the nation's electric power. That said, on a global scale wind is still a bit player, cornering only 0.31 percent of the total electricity market.

Despite its near price parity, wind power suffers from an important disadvantage that precludes it from ever becoming the dominant electricity source for the industrial world. The problem is that the wind is intermittent (as is sunlight), and so it requires another, more reliable form of power—such as fossil fuel, hydroelectric, or nuclear—to also be available to take up the slack when the wind drops. Since the power produced by a wind turbine goes as the cube of the air speed, if the wind drops to half the turbine's design value, its output is cut by a factor of eight—and wind sometimes dies completely. So even if it possessed all the wind turbines it could ever want, a nation would still need to have enough conventional power capacity to support 100 percent of its demand. The same disability would also be incurred by a solar power generation system, but at much higher cost.

That said, there are certain applications where wind has advantages. It is very good for use in isolated locations, such as third world villages, where logistic supply of fuels may be difficult, and where power demand is not that large. A wind turbine with intermittent output can still be used to recharge batteries, allowing rural villagers to make use of transistor radios, cell phones, and even laptop computers, thereby linking them to the world at large. Even a few electric lights can radically transform people's lives by making it possible for them to read at night. Most important, the windmill's intermittent power output is not a major problem if its assigned task is to pump water for irrigation purposes. Historically, this has been a very important role for windmills in rural areas.

Growing plants is a great way to fight global warming, both because they consume carbon dioxide and because the evaporation of water from their leafy surfaces cools the land. If you want to see how strong an effect that is, just compare the summer weather reports in Birmingham, Alabama, and Baghdad, Iraq. Both cities are at exactly the same latitude (33° 25' N) but the daytime highs in Baghdad average 11°C (20°F) hotter. That's the handiwork of vegetation. In the future we may want to set up coastal installations that desalinate seawater (an application for which simple solar heating reflectors—but not photovoltaics—are well suited) and then pump the freshwater product inland to make the deserts bloom. Such valuable work can well be done by windmills.

But if we want meet the massive power needs of growing industrial societies with a zero-emission electricity source, there is only one way to do it. We are going to need to use nuclear power.

There are two kinds of nuclear power: fission and fusion. Fission works by using a neutron as a kind of projectile to split the nuclei of very heavy atoms such as uranium or plutonium into middle-weight elements, thereby releasing energy as well as several more neutrons that can be used to continue to process in a chain reaction. Fusion, which is the reaction that powers the sun and all the stars, works by fusing the nuclei of hydrogen isotopes into helium, in the process releasing even more energy than fission. Fission reactors have been a practical means of generating electricity at commercially competitive rates since the 1950s. Controlled fusion is still experimental.

Despite its lack of air pollution or greenhouse emissions, nuclear fission has provoked much opposition from environmentalists concerned about the possibility of nuclear accidents as well as the problem of disposing of its radioactive waste products. These concerns, however, have been largely misplaced.

Over the course of its entire history, the world's commercial nuclear industry has had two major accidents, one at Three Mile Island in Pennsylvania in 1979, and the other at Chernobyl in the Ukraine in 1986. The Three Mile Island event was a core meltdown caused by a failure of the cooling system. A billion-dollar reactor was lost, but the containment system worked. As a result, there were no human fatalities or any significant environmental impact.

Chernobyl was a much more serious affair. At Chernobyl, the reactor actually had a runaway chain reaction and exploded, breaching all containment. Approximately fifty people were killed during the event itself and the fire-fighting efforts that followed immediately thereafter. Furthermore, a radiological inventory comparable to that produced by an atomic bomb was released into the environment. According to an authoritative twenty-year study by the International Atomic Energy Agency and the World Health Organization,[6] over time this fallout will most likely be responsible for four thousand deaths among the surrounding population. So Chernobyl was really bad. Yet in comparison to the *millions* of deaths caused every year as a result of the pollution emitted from coal-fired power plants, its impact was minor. It would take a Chernobyl event *every day* to induce a casualty rate comparable to that currently being inflicted on humanity by coal. By replacing a substantial fraction of the electricity that would otherwise have to be generated by fossil fuels, the nuclear industry has actually saved tens of millions of lives.

Still, events like Chernobyl need to be prevented, and they can be—by proper reactor engineering. The key is to design the reactor in such a way so that as its temperature increases, its power level will go *down*. In technical parlance, this is known as having a "negative temperature coefficient of reactivity." As early as 1950, Captain Hyman Rickover, the leader of the effort to create the first nuclear submarine, realized that having such intrinsic system negative feedback against power spikes was fundamental to ensuring safe operation of a practical nuclear reactor. Accordingly, the reactor of the *Nautilus* was designed in such a way that a chain reaction could not be sustained unless liquid water was present in the cooling channels throughout its core. From a nuclear point of view, the water is a necessary ingredient in such a system, because it serves to slow down, or "moderate," the fast neutrons born of fission events enough for them to interact with surrounding nuclei to continue the reaction. (Just like an asteroid passing by Earth, a neutron is more likely to be pulled in to collide with a nucleus if it is moving slowly than if it is moving quickly.) It is physically impossible for such a water-moderated reactor to have a runaway chain reaction, because as soon as the reactor heats beyond a certain point, the water starts to boil. This reduces the water's effectiveness as a moderator, and without moderation, fewer and fewer neutrons strike their target, causing the reactor's power level to drop. The system is thus intrinsically stable, and there is no way to make it unstable. No matter how incompetent, crazy, or malicious the operators of a water-moderated reactor might be, they can't make it go Chernobyl. It was on this principle that Rickover designed both the *Nautilus* and the subsequent first civilian nuclear reactor at Shippingport, Pennsylvania, and these have served to set the pattern for nearly all American reactors ever since.

In contrast, the Soviet RBMK reactor that exploded at Chernobyl was moderated not by water, but by graphite, which does not boil. It therefore did not have the strong negative temperature reactivity feedback of a water-moderated system, and in fact, due to various other design features, it actually had a positive temperature coefficient of reactivity. It was thus an unstable system, and could be—and was—set off when its operators decided to do some really dumb experiments. No such system could ever be licensed in the United States.

Airplanes can crash, and bridges and apartment buildings can collapse, if they are poorly designed. In this sense, nuclear power is no different from any other complex engineering system. It needs to be done right. But it can be.

As for the issue of nuclear waste disposal, it has been wildly overdrawn by ideological antitechnology organizations seeking to create a showstopper for the nuclear industry. Such groups claim to be interested in public safety and environmental preservation. Yet it must perplex the rational mind that anyone can agitate, litigate, and argue with a straight face that it is better that nuclear waste be stored in hundreds of cooling ponds adjacent to reactors located near metropolitan areas all across the country, rather than gathered up and laid to rest in a government-supervised desert depository deep under Nevada's Yucca Mountain. Clearly, such people are more interested in manufacturing a problem than in solving one.

The Department of Energy's (DOE) Yucca Mountain plan is perfectly adequate to ensure public safety. My only objection to it is that it is economically wasteful. A more cost-effective solution would be to simply glassify the waste into a water-insoluble form, put it in stainless steel cans, take it out in a ship, and drop it into midocean subseabed sediments that have been, and will continue to be, geologically stable for tens of millions of years. Falling down through several thousand meters of water, such canisters readily reach velocities that allow them to bury themselves hundreds of meters under the mud. After that, the waste is not going anywhere, and no one, by accident or design, will ever stumble upon it.

This solution has been well known for years,[7] but has been shunned by DOE bureaucrats who prefer a large, land-based facility because it involves a much bigger budget. Nevertheless, the Yucca Mountain plan will work.

Another concern regarding nuclear power is its potential to act as an agent for atomic weapons proliferation. Natural uranium contains 0.7 percent uranium-235 (U-235), which is fissile, and 99.3 percent uranium-238 (U-238), which is not. In order to be useful in a commercial nuclear reactor, the U-235 fraction is typically enriched to 3 percent concentration. The same enrichment facilities could also be used, with some difficulty, to further concentrate the U-235 to 93 percent, which would make it bomb-grade. Also, once placed in the reactor, some of the U-238 will absorb neutrons, transforming it into plutonium-239 (Pu-239), which is fissile. Ideally, such plutonium would be reprocessed out of the spent fuel and mixed with natural uranium to turn it into reactor-grade material, but it could also be used to make bombs instead. Thus the technical infrastructure required to support an end-to-end nuclear industry fuel cycle could also be used to make weapons.

This, however, does not mean that nuclear power plants should be avoided. It is the enrichment and reprocessing facilities that actually present the weapons-making danger, and such capabilities should certainly be forbidden to terrorist nations like Iran. But the power stations themselves are not the threat. If plutonium is desired, much better material for weapons purposes can be made in stand-alone atomic piles divorced from any other function than in commercial power stations.[8] Both the United States and the Soviet Union had thousands of atomic weapons before either had a single nuclear power plant, and others desirous of obtaining atomic bombs can proceed the same way today.

That said, nuclear power is not for every country. Nuclear power plants are very expensive. They require an educated workforce, a responsible government, and a safe security environment. It would be insane, for example, to build a nuclear power plant today in the Congo or Somalia. That would be like putting an original Tang dynasty vase in the middle of a kindergarten playground. However, as it turns out, for logical and necessary reasons, the largest consumers of electricity worldwide are well-organized countries. Thus nuclear power will be able to be brought to bear to meet most of the needs of an advancing world. As for those nations that are not yet ready for nuclear power, the task of our age is to help transform them so that they are. The issue is not one of "inappropriate technology," but inappropriate social conditions. Backwardness, savagery, and anarchy are the things that are inappropriate. For many reasons that go well beyond energy policy, we need to make the world safe for nuclear power.

THE POWER THAT LIGHTS THE STARS

Nuclear fission has radically expanded humanity's potential energy resources. The achievement of controlled nuclear fusion will make them virtually infinite. The basic fuel for fusion is deuterium, which is an isotope of hydrogen that is called "heavy" because, in addition to having the standard proton in its nucleus that all hydrogens have, it also has a neutron, which doubles its weight. Deuterium is found naturally on Earth; one hydrogen atom out of every six thousand is deuterium. That might not seem like much, but because of the enormous energy released when a fusion reaction occurs, it's enough to endow each and every gallon of water on Earth, fresh or salt, with a fusion energy content equivalent to that obtained by burning 350 gallons of gasoline. To see what this means for the human future, take a look at the comparison of Earth's energy resources listed in table 11.2, where the resource size is given in terms of terawatt (TW = trillion watt) years. For comparison, currently humanity collectively uses about 13 TW of power. Earth's fusion resources are more than a million times greater than all other energy reserves put together. Even at ten times our current rate of consumption, there is enough fusion fuel on this planet (alone) to power our civilization for nearly a billion years.

Furthermore, fusion produces no greenhouse gases, and if done correctly, need not produce significant radioactive waste. When they collide, the deuterium nuclei fuse to form tritium or helium-3 (He-3) nuclei, plus some neutrons. The tritium and He-3 will then react with other deuteriums to produce ordinary helium (He-4) and common hydrogen (H-1), plus a few more neutrons. If the reactor is made of

TABLE 11.2. THE EARTH'S ENERGY RESOURCES[9]

Resource	Energy (TW-years)
Oil (known reserves)	202
Coal (known reserves)	790
Natural Gas (known conventional reserves)	205
Natural Gas (including sub-sea methane hydrates)	24,000
Nuclear Fission (commercial grade U ore, without reprocessing)	685
Nuclear Fission (commercial grade U ore, with fuel reprocessing)	50,000
Fusion	100,000,000,000

conventional materials, like stainless steel, the neutrons can produce some activation, resulting in the production of about 1/1,000th the radioactive waste of a fission reactor. However, if specially chosen structural materials like carbon-carbon graphite are used, there will be no activation, and the system can produce endless amounts of energy without pollution of any kind.[10]

Once we have fusion, we will be able to make as much methanol as we desire inorganically, simply by reacting carbon dioxide with electrolysis-produced hydrogen over copper on zinc oxide catalyst. The carbon dioxide could come from high carbon dioxide emission sources such as steel mills or cement plants, or even directly from the atmosphere itself. Under such circumstances, the Islamists' possession of the world's oil reserves will give them as much influence over the human future as they currently derive from their monopoly of camel milk.

But fusion is not just a plentiful source of energy; it is a new *kind* of energy, one that offers the potential to do things that are simply impossible without it. If we can get fusion, we will be able to use the superhot plasma that fusion reactors create as a torch to flash any kind of rock, scrap, or waste into its constituent elements, which could then be separated and turned into useful materials. Such technology would eliminate any possibility of resource exhaustion of this planet. Using fusion power, we will be able create space propulsion systems with exhaust velocities up to *five thousand times* greater than the best possible chemical rocket engines. With such technology, the stars would be within our reach.

So the fusion game is really worth the candle. It's a tough game, though, because while fusion occurs naturally in the stars, creating the conditions on Earth to allow it to proceed in a controlled way in a human-engineered machine is quite a challenge.

All atomic nuclei are positively charged, and therefore repel each other. In order to overcome this repulsion and get nuclei to fuse, they must be made to move very fast while being held in a confined area where they will have a high probability of colliding at high speed. Superheating fusion fuel to temperatures of about 100 million °C gets the nuclei racing about at enormous speed. This is much too hot to confine the fuel using a solid chamber wall—any known or conceivable solid material would vaporize instantly if brought to such a temperature. However, at temperatures above 100,000°C, gases transition into a fourth state of matter, known as a plasma, in which the electrons and nuclei of atoms move independently of each other. (In school we are taught that there are three states of matter: solid, liquid, and gas. These dominate on Earth, where plasma exists only in transient forms in flames and lightning. However, most matter in the universe is plasma, which constitutes the substance of the sun and all the stars.) Because the particles of plasma are electrically charged, their motion can be affected by magnetic fields. Thus magnetic traps such as the toroidal or doughnut-shaped tokamak (as well as a variety of alternate concepts including stellarators, magnetic mirrors, etc.) have been designed that can contain fusion plasmas without ever letting them touch the chamber wall.

At least that is how it is supposed to work in principle. In practice, all magnetic fusion confinement traps are leaky, allowing the plasma to gradually escape by diffusion. When the plasma particles escape, they quickly hit the wall and are cooled to its (by fusion standards) very low temperature, thereby causing the plasma to lose energy. However, if the plasma is producing energy through fusion reactions faster than it is losing it through leakage, it can keep itself hot and maintain itself as a standing, energy-producing fusion fire for as long as additional fuel is fed into the system. The denser a plasma is, and the higher its temperature, the faster it will produce fusion reactions, while the longer individual particles remain trapped, the slower the rate of energy leakage will be. Thus, the critical parameter affecting the performance of fusion systems is the product of the plasma density (in particles per cubic meter), the temperature (in kilovolts, or keV, 1 keV = 11,000,000°C = 20,000,000°F), and the average particle confinement time (in seconds) achieved in a given machine. The progress that the world's fusion programs have had in raising this triple product, known as the Lawson parameter, is shown in figure 11.1.

The easiest fusion reaction to drive is that between deuterium and tritium. To produce energy at a rate equal to the external power being used to heat the plasma (via microwave heaters or other means), a deuterium-tritium (D-T) fusion plasma must have a Lawson parameter of 9×10^{20} keV-particle-seconds/m^3 (or keV-s/m^3 for short). Such a condition is known as "breakeven" and was finally reached at the European JET tokamak in 1995. A deuterium-tritium plasma with a Lawson parameter of 4×10^{21} keV-s/m^3 (the notation 4×10^{21} means a 4 with 21 zeroes after it) and a temperature of 10 keV would produce energy at a sufficient rate that no external heating power would be needed. Once started up, such a plasma would heat itself. This condition is known as "ignition," and is the next, and final, major physics milestone that needs to be achieved before actual energy-producing fusion reactors can be engineered.

A fusion reactor could be operated as a D-T system, obtaining its tritium by reacting the neutrons it emits with a lithium blanket surrounding the reactor vessel. (When a lithium nucleus absorbs a neutron it splits into a helium and a tritium, and sometimes emits a neutron, which allows yet another tritium to be produced.) First-generation fusion reactors may be designed along these lines.[11] However, with a little further progress in improved magnetic confinement, this will become unnecessary. Instead, once ignition is reached, we will be able to use the plasma's own power to ramp its temperature up to 40 keV, at which point the deuterium and its by-products will burn by themselves without engineered tritium enrichment.[12]

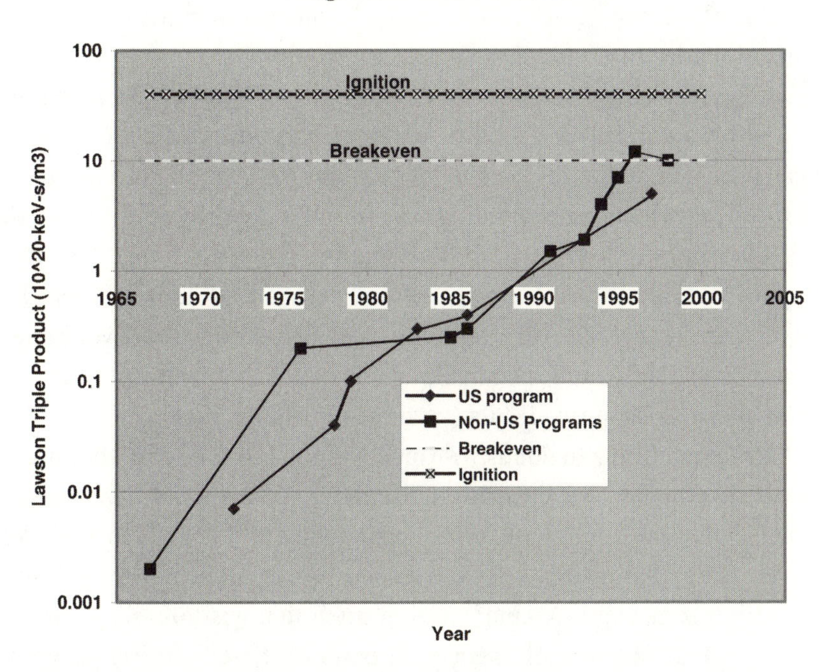

Figure 11.1. Progress in controlled fusion. Since 1965, the world's fusion programs have advanced the achieved Lawson parameter by a factor of ten thousand. A further increase of a factor of four will take us to ignition. Note the logarithmic scale. *Source*: Adapted from data provided by European Fusion Development Agreement, www.efda.org.

As can be seen in figure 11.1, the world's fusion programs have made an enormous amount of progress over the past thirty years, raising the achieved Lawson parameter by a factor of almost ten thousand to reach breakeven. Another factor of four, which can certainly be accomplished if funds are provided to build the next generation of experimental tokamaks, would take us to ignition.

Fusion can certainly be developed, and when it is, it will eliminate the specter of energy shortages for millions of years to come.

WRECKING THE FUTURE

Research in controlled fusion has been under way since the 1950s. Why has it not yet achieved success? The answer is a gross failure of political leadership.

In figure 11.2 we show the US fusion budget from 1953 to the present, in both current-year dollars and inflation-adjusted 2006 dollars. It can be seen that the fusion budget was kept very small until the 1973 oil shock, at which point funding was raised to healthy levels of $800 to $900 million per year (in 2006 dollars). Under the impetus of this support from the Ford, Carter, and first Reagan administrations, the fusion program took off, building and operating the ever-better machines that delivered spectacular gains in achievement from the late 1970s through the early 1990s (as shown in figure 11.1).

However, in 1986 OPEC dropped the oil price, and the smart people responsible for ensuring US energy security had a very deep insight. They said, "Oh look, oil is cheap today. Who needs fusion? We can get all of the cheap energy we need just by depending upon our good friends in Saudi Arabia." Acting upon this wisdom, they cut the budget in half. As a result, nearly all parts of the US fusion program except for the tokamak were rapidly shut down. (In one particular atrocity, an extremely important experiment, the Livermore Lab MFTF-B magnetic mirror, built over four years at a cost of some $400 million in today's money, was decommissioned the same month it was completed, with scientists never even given a chance to turn the machine on and see what it could do.)

The surviving tokamak program was allowed to limp into the 1990s, making progress using the Princeton Plasma Physics Lab TFTR machine built with Carter and early Reagan administration money. But then, in 1996, the inestimable Mr. Gore, leading the Clinton administration's science policy, showed the sincerity of his commitment to find constructive solutions to global warming and cut the fusion budget nearly in half again. As a result of these cuts (which were also enthusiastically supported by foolish fiscal conservatives in Congress), plans to build a follow-on machine to the aging TFTR to advance its work to ignition had to be aborted, and the whole program was put on idle.

If you want to get some idea of the mindlessness of the decision-making process of the American political class with respect to the fusion program, take a look at figure 11.3, which compares the history of US fusion expenditures to the variation of the price of crude oil.

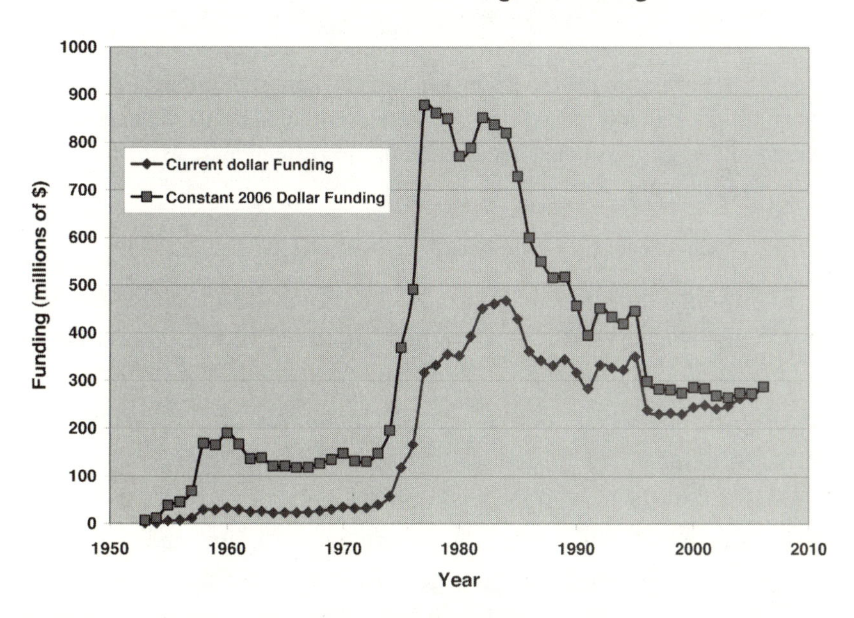

Figure 11.2. US Fusion Program Funding, 1953–present. *Source*: Data courtesy of Fusion Power Associates, www.fusionpower.org.

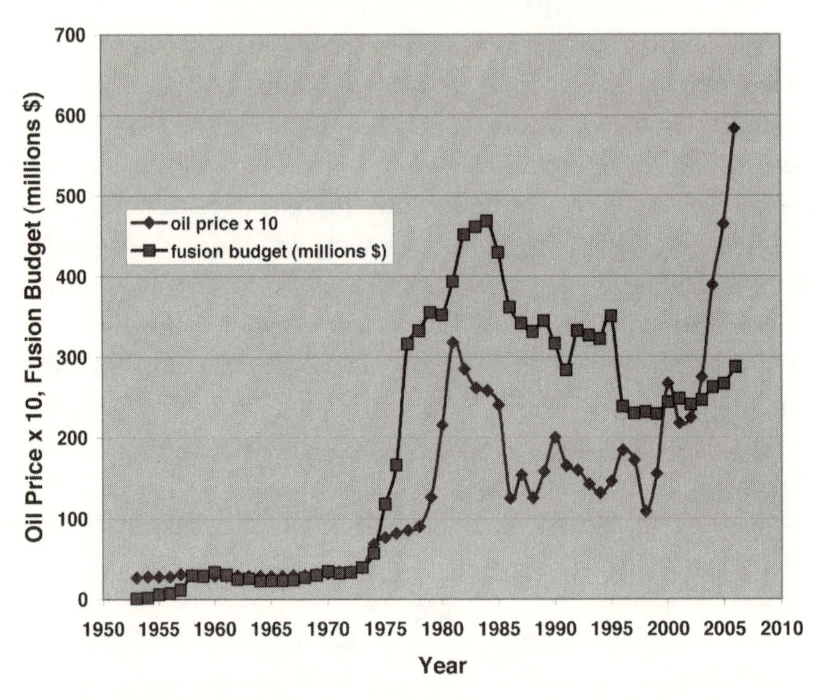

Figure 11.3. Comparison of the US fusion budget with the oil price, in current dollars. Oil price shown is the price per barrel, multiplied by 10. *Source*: Data courtesy of Energy Information Administration and Fusion Power Associates, www.fusionpower.org.

Examining the data in figure 11.3, you can see that for the past forty years, the yearly fusion funding approved by the US Congress has directly tracked the price of oil. The foolishness that this shows is amazing. Effectively, America's yearly fusion research appropriations are being determined by the whims of OPEC. We cannot conduct a successful long-term energy development program on such a basis.

In constant-dollar terms, the US fusion program today receives less than one-third the funding it got in 1980. As a fraction of the federal budget, it is receiving under one-eighth its former support. The US fusion program's total nationwide annual budget is less than one-quarter of what NASA spends on a single space shuttle launch. With funding levels kept so low, progress is nearly impossible.

But, as harmful as it has been, the feckless denial of necessary funding has not been the only form of mismanagement stifling the fusion program. The other has been a decision to turn the program into a diplomatic plaything under the banner of "international collaboration."

Up through the mid- to late 1980s, the worldwide fusion research field was a vigorously competitive arena, with various national fusion programs vying to outdo and upstage each other with surpassing achievements. The atmosphere was like a race. At every major conference, the Americans, the Russians, the Europeans, and the Japanese would be there, trying to make a splash with their latest results, or astounding the rest by making public announcements of plans for their next bold steps. If people in the West wanted to be complacent about fusion, tough competitors like the brilliant Soviet scientists Andrei Sakharov and Lev Artsimovich were quite ready to shake them up.[13] Thus, for example, it was the announcement of spectacular advances by Artsimovich and his Kurchatov Institute T-3 tokamak team in 1968 that Sputniked the West into abandoning some well-dug-in approaches that were going nowhere and launching its own tokamak program to catch up.

In the mid-1980s, however, a number of bureaucrats in charge of the American, Soviet, European, and Japanese fusion programs conceived the idea that it would be better to end all this messy and expensive competition and instead merge all the world's fusion efforts into a single unified program. So, discarding previously made plans for aggressive furthering of their own national programs, they came together to launch the International Thermonuclear Experimental Reactor (ITER) project. According to the new plan, humanity would pool its research efforts in ITER and take its further steps toward controlled fusion energy in conjoined collaboration.

In the two decades since its initiation, the ITER program has held scores of "working group" meetings in various prestigious or scenic locations around the world, affording those involved with excellent opportunities to rack up frequent-flier airline miles. Reportedly, the food served at many of the banquets has been excellent, too. The productivity of these meetings has been remarkable. In 2005, for example, after twenty years of deliberations, the ITER committee actually decided where the experiment will be placed (France). With this important milestone behind it, the ITER leadership committee is now positioned to move ahead rapidly. Indeed, it is rumored that in just a few more years, ground may actually be broken to begin construction. In the meantime, several more top-level meetings are planned.

In short, ITER has been a poster child for the argument that in science, as in all other human endeavors, competition is an essential spur for progress. Instead of creating a mighty united alliance to advance worldwide fusion research, ITER has served as a kind of international suicide pact, killing motivation and drive, and enabling mediocrity—or rather, complete program stagnation—to become the mutually accepted standard of accomplishment.

In 2002 a number of laboratory leaders from what was left of the US fusion program saw the writing on the wall and decided to pull together behind a proposal to launch a new American effort separate from the ITER bureaucratic quagmire. The concept was that the United States would take the initiative (imagine that!) to build its own next-generation tokamak fusion machine, called the Fusion Ignition Reactor Experiment (FIRE). In contrast to ITER, whose design was based upon an international consensus to simply scale up a 1980s-style tokamak in size to improve confinement time (which should improve with the square of the reactor's characteristic dimension), FIRE would take the more adventurous approach of increasing *both* confinement time and plasma density simultaneously. This could be done without increasing the size of the machine at all by ramping up the power of the tokamak's magnetic field. Indeed, since both density and confinement time increase in proportion to the square of the magnetic field strength, doubling the field would allow FIRE to achieve ignition with a machine one-quarter the size of ITER.[14] Because its configuration is so much smaller, FIRE would be an order of magnitude cheaper than ITER, and could be easily afforded by America acting alone. It could also be built much more quickly, and thus provide real experimental data sooner, which then could be used to provide the basis for the design of still more advanced machines. Finally, as a number of FIRE advocates have wryly noted, the existence of an independent US fusion program would help, rather than harm, the international effort, by "providing ITER with a useful yardstick against which to measure its own progress." (Translation: FIRE would give ITER a vitally necessary kick in the butt.)

The fortunes of FIRE have waned and waxed since its conception. As I write these lines, the question of whether or not it will be funded is being fought out among various factions in the Department of Energy and the US Congress. Let's hope it wins, through, because unless something like FIRE is started soon, the US fusion program will totally collapse and ITER will stagnate itself to death. Were that to happen, the knowledge and know-how gained from the past fifty years of worldwide fusion research would largely be lost. That is a result we cannot accept, because fusion is vital to a hopeful human future.

OUR CHOICE OF FUTURES

Humanity today stands at a crossroads, facing a choice between two very different visions of the future. On the one side stand those who postulate a world of limited supplies, whose fixed constraints demand ever-tighter controls upon human aspirations. On the other stand those who believe in the power of unfettered creativity to invent infinite resources, and so, rather than regret human freedom, insist upon it.

Fusion power is many things. It is, first and foremost, a practical means of providing endless, nonpolluting energy to sustain an ever-progressing human civilization on this planet for millions of years to come. It is also, I am firmly convinced, the power that will allow our descendants to settle other planets and, someday, travel to the stars.[15] And while the actualization of such possibilities will be the joyful work of future ages, the way we view them from afar will very much affect the nature of our own.

If the idea is accepted that the world's resources are fixed with only so much to go around, then each new life is unwelcome, each unregulated act or thought is a menace, every person is fundamentally the enemy of every other person, and each race or nation is the enemy of every other race or nation. The ultimate outcome of such a world-view can only be enforced stagnation, tyranny, war, and genocide. Only in a world of unlimited resources can all men be brothers.

That is why, even though we have enough other energy supplies to sustain our society for quite a while, we should invest the effort to develop fusion power now. For, in doing so, we make a statement that we are living not at the end of history, but at the beginning of history: that we believe in freedom, and not regimentation; in progress, and not stasis; in love, rather than hate; in life, rather than death; in hope, rather than despair.

CHAPTER 12
OIL AND POWER

> The use of oil made it possible in every type of vessel to have more gunpower and more speed for less size or less cost. It alone made it possible to realize the high speeds in certain types which were vital to their tactical purpose. . . . But oil was not found in appreciable quantities in our islands. If we required it, we must carry it by sea in peace or war from distant countries. . . . To commit the Navy irrevocably to oil was indeed 'to take up arms against a sea of troubles.' . . . Yet beyond the breakers was a great hope. If we overcame the difficulties and surmounted the risks, we should be able to raise the whole power and efficiency of the Navy to a definitely higher level; better ships, better crews, higher economics, more intense forms of war power—in a word, mastery itself was the prize of the venture.
>
> —Winston Churchill, *The World Crisis*[1]

The oil industry involves a great deal of money, but fundamentally it is not about money; it is about power. For nearly a century, control of oil has been the decisive factor determining victory or defeat in the struggle of nations for world dominance. If we want to win the war on terror, we need to fully understand this very critical point.

It was in the first days of World War I that the potential military significance of oil first became apparent. The war had begun in early August 1914 following the script drawn up by the German general staff, with the French, as expected, launching the bulk of their army eastward in a futile direct assault to try to recover their beloved Alsace-Lorraine. Anticipating this move, the Germans had left just one army in those provinces to hold the French off, while sending six other armies through neutral Belgium to sweep around through northern France, and then turn east to take the main French forces in the rear. The strategy, known as the Schlieffen plan, was modeled on that employed by the Carthaginian general Hannibal against the Romans at Cannae in 212 BCE. On that battlefield, Hannibal had outflanked and enveloped sixteen legions, massacring eighty thousand Romans in a single day. The Schlieffen plan was anticipated to do the same thing on a vastly expanded scale, surrounding and slaughtering more than a million French soldiers in a colossal "super-Cannae."

Despite the fact that they knew all about the Schlieffen plan, the French high command blundered right into it. By early September the Germans marching down from Belgium had reached the Seine, and were driving the French eastward toward the border, where they would be caught in a giant pocket and annihilated. Both sides were nearly exhausted from weeks of nonstop marching and fighting, but the Germans were better officered and far better disciplined. Punch-drunk and sleepless though they were, they relentlessly slogged forward, driving the French toward doom.

There was, however, one French division that was not caught in the trap. It was the garrison of Paris, commanded by the ancient General Joseph Gallieni. A veteran of the Franco-Prussian War of 1870–71, the impossible General Gallieni had refused the desperate pleas of the French high command to send his force to join the rest of the army in the main battle. Instead, anticipating another siege like that of 1870, he insisted on keeping his division in Paris to defend the city.

The Germans, however, did not attack Paris. Their target was the French field army. Once that was destroyed, they could deal with the capital and its garrison at their leisure. With singleness of purpose, they swept past the city to assault the retreating French, who had turned to try to make a weary last stand on the river Marne. Gallieni's troops were left behind in Paris with no one to fight.

Gallieni, however, was not quite as senile as he appeared. He saw that events had now placed his force to the German rear. If only there was some way he could reach the battle, he could hit the enemy from behind. But how? The war had made the usually chaotic French railroad system completely nonfunctional, and the Germans were forty miles away. For his infantry to reach the battle by foot would require a forced march of two days. By then it would too late.

The situation seemed hopeless, but then, during the evening of September 6, 1914, Gallieni had stroke of genius. Use automobiles! Cars were not in general use in France in 1914, but there were several thousand taxicabs in Paris—*enough to move a division.* He put out the word to the cabbies of Paris that he would pay them on the meter if they would drive his army to the Marne. Within two hours, dozens of cabs had gathered at the designated assembly point at the Esplanade des Invalides. By morning there were over six hundred. Piling onboard, the men of Gallieni's "taxicab army" set off in a huge convoy down the Champs-Elysees and reached the Marne within hours. Arriving without warning on the exposed German flank, Gallieni's fresh troops launched a furious attack that sent the kaiser's tired army reeling into pell-mell retreat. It was the first motorized infantry assault in history, and it changed the outcome of the war.

Observing his valiant force as it wheeled out of Paris toward its rendezvous with destiny, Gallieni is reported to have remarked to one of his aides, "Well, this at least is something out of the ordinary."[2] But as the war went on, the role of mechanized transport changed from a brilliant improvisation into a daily essential of military logistics. The numbers tell the tale. In 1914, for example, the British Expeditionary Force in France had a total of fewer than 850 motorcars, and no trucks. By 1918 they had 56,000 trucks, 23,000 motorcars, and 34,000 motorcycles.[3] The French grew their mechanized transport arm in comparable fashion, and when the Americans showed up, they brought yet another 50,000 trucks and cars. It was only through the use of these vast numbers of vehicles that the Allies were able to shift their forces from one sector of the front to another rapidly enough to counter the troop movement capability offered by the far superior German railway network.

At the Battle of the Somme in 1916, the British introduced the tank to warfare, and the role of the internal combustion engine expanded from transport to combat. Two years later tanks were dominating the battlefield, and when, on August 8, 1918, a force of 456 Allied tanks broke the German line at the Battle of Amiens, the four-year stalemate on the western front came to an end.

An even more revolutionary oil-powered military development occurred above the battlefield. When the war began, airplanes were a curiosity. By its end, the Allies had produced and pressed into service more than 158,000 aircraft—and the Germans, 48,000—with quality improving by the month. In the battles of 1918, the Allied air superiority was used to powerful effect in tactical support of ground operations on the western front, and achieved devastating results against the Turks in the Middle Eastern theater.

OIL RULES THE WAVES

Yet as important as its influence in land and air combat may have been, it was at sea that oil demonstrated its decisive strategic importance. In contrast to coal, oil, because it enabled refueling at sea, could give propeller-driven warships true global reach. The visionary British admiral John Arbuthnot "Jacky" Fisher had understood this fact before the war, and impressed it upon his protégé, the young Winston Churchill. Consequently, when Churchill became First Lord of the Admiralty in September 1911, he launched a crash program to convert the British navy from coal to oil.[4]

It was this decision by Fisher and Churchill that transformed oil from simply one commodity among many into the single strategic resource determining the outcome of the struggle for world dominance. For centuries world power had depended upon sea power, because whoever controlled the seas controlled access to all the products and wealth of the globe. This principle, well understood by the English power elite, was the central foundation of the British Empire. In abandoning coal, the Fisher-Churchill policy turned the British navy back into the true blue-ocean globe-spanning force it had been in the age of fighting sail. But henceforth, British sea power, and thus the very existence of the empire itself, would depend critically upon oil.

Prior to the war, the Germans had also built up an impressive battleship navy. In numbers the kaiser's dreadnoughts were nearly a match for the British Grand Fleet, and in armor, armament, and general quality of construction they were better. But they had a critical weakness: They were powered by coal, and that, combined with Germany's lack of overseas coaling stations, effectively meant they couldn't go anywhere. During the entire war, the pompously misnamed German "High Seas Fleet" never left the North Sea.

The Germans, however, did have one naval arm that could effectively challenge the British on the open ocean, and that was their fleet of wide-ranging diesel-powered submarines. While east of Suez, the British navy could obtain its fuel from its Anglo-Persian Oil Company (now called British Petroleum) concessions, but the British home fleet was completely dependent upon oil shipments from the United States. Indeed, in 1914 the United States was responsible for 67 percent of all oil production worldwide. If the German U-boats could cut the transatlantic oil lifeline, the British navy would be immobilized. Under those conditions, the British Isles could be blockaded into starvation, and with their cross-channel supplies cut off, the French would face defeat as well.

During 1915 and 1916 the U-boats sank a great deal of Allied shipping, creating what the London *Times* called "a dearth of petrol" in Britain and France. But so long as neutral (i.e., American) tankers could ply their trade unimpeded, enough oil would still come through to allow the Allies to get by. If the Allies were to be brought to their knees through a fuel cutoff, the American tankers would have to be attacked as well. Viewing the problem this way, the Germans took their greatest gamble. In the spring of 1917 they announced a policy of unrestricted submarine warfare, targeting Allied and neutral shipping alike.

The decision has been called an insane blunder, because it brought the United States into the war on the Allied side. But it nearly worked. In the first few months after the U-boats were unleashed, Allied (now including American) tanker losses were catastrophic. By May 1917, Royal Navy fuel reserves were down to less than a three months' supply, and necessary operations were being curtailed. By summer the situation was desperate. "The Germans are succeeding," American ambassador Walter H. Page wrote in July. "They have lately sunk so many oil ships that this country may very soon be in a perilous condition—even the Grand Fleet may not have enough fuel. . . . It is a very great danger." Gasoline rationing was progressively tightened until, in October 1917, pleasure driving in Britain was banned completely. In December French prime minister Georges Clemenceau made a desperate plea to President Woodrow Wilson to get more oil through quickly, whatever the cost. With the Russians knocked out of the war, the Germans were shifting troops from the eastern front to the west, and a massive offensive was expected shortly. Without gasoline for the trucks needed to move their troops, the French would be helpless. "Gasoline is as vital as blood in the coming battles," Clemenceau declared.[5]

It was barely in the nick of time that the Allies developed an adequate convoy system to stop the U-boats. In the spring of 1918, transatlantic oil tankers started arriving again in increased numbers, and with them came the fresh-faced boys of the American Expeditionary Force. Refueled and reinforced, the Allies stopped Erich Ludendorff's last great summer offensive, and then, taking advantage of their superiority in tanks, trucks, and aircraft, pushed the Germans back. The crack German infantry were as tough as they come, but there was no way they could cope with a new army equipped with fleets of rampaging gasoline-powered land battleships and assisted by unmatched swarms of fighter aircraft. Abandoning one trench line after another, they began a grim retreat that did not stop until Germany surrendered a few months later.

At a victory banquet in London on November 18, 1918, Lord Curzon declared, "The Allied cause had floated to victory upon a wave of oil."[6]

THE NAZIS PREPARE FOR WAR

In many ways, the First World War on the western front had been a contest between two technological systems, with coal, steam, and railroads facing off against oil, internal combustion engines, and automotive transport. Having virtually no oil resources of their own, the Germans had opted for coal—and lost. And while events might have turned out differently had alternative tactical decisions been made at various points, it was clear enough by the time it was over in what direction the future lay. Thus, as early as 1921 the management of the I. G. Farben Corporation, foreseeing another war as inevitable, bought up the patents for a process invented by the chemist Frederic Bergius in 1909 for producing synthetic gasoline. The Bergius process, which involved reacting hydrogen with coal at high temperature and pressure, yielded a high-quality product, but at six times the cost of gasoline manufactured from natural petroleum. Accordingly, it could be implemented only with the help of massive government subsidies. In the next war, however, gasoline would be an absolute necessity; to ensure an adequate supply, Farben management argued, no price could be too high. In the wake of Germany's bitter defeat, their case found many supporters in the German general staff, and they were able to convince the conservative Weimar government to provide the funds necessary to get a synthetic fuel industry started.

Because about a third of the Farben board was Jewish, the Nazis regarded the Weimar government's synthetic fuel subsidies to the company as a major scandal (German taxpayer money going to Jewish plutocrats, etc.) and denounced the program regularly in their propaganda. However, as soon as he took power in 1933, Adolf Hitler realized the importance of the initiative for his own plans for world conquest. So rather than shut the subsidies off, he purged the company of its Jewish and other anti-Nazi elements, and then radically expanded the program.[7]

On a technical level, the results of the synthetic fuel program were impressive. Huge production works were built at Leuna in east-central Germany that by 1939 were making over 33 percent of Germany's gasoline, a figure that rose to 75 percent, or 7 million tons per year, by 1943. The quality of the product was high, so much so that virtually all of the high-octane aviation fuel Germany used during the war was made at the Farben Leuna complex.

Additional supplies of oil were obtained from the Ploesti petroleum fields in Axis-allied Romania, and for a while, under the 1939 Hitler-Stalin Pact, from the Soviet Union. By combining these somewhat limited sources with his synthetic gasoline capability, Hitler was able to fuel the Panzer divisions and armadas of bombers that blitzed Poland, France, Scandinavia, the Low Countries, and Britain during 1939 and 1940. In June 1941, however, the Nazis invaded the Soviet Union, and Germany's fuel requirements expanded radically.

While keenly aware of the logistic dimension of warfare, Hitler nevertheless believed that the quick collapse of Soviet resistance would make the narrowness of his fuel base irrelevant. However, when the Russians defied expectations and rallied to halt the German advance before Moscow in December 1941; the Nazis found themselves facing the prospect of a sustained continental-scale land war against an opponent who was becoming stronger, smarter, and more determined by the month. And if the destruction wrought upon the frozen German lines during the winter of 1941–42 by mounted Cossack raiders wasn't bad enough, in the spring of 1942 the Russians started to bring thousands of excellent T-34 tanks into action. If something wasn't done, the Germans would soon find themselves on the receiving end of a blitzkrieg.

The issue hinged upon oil. On the vast Russian steppe, the tank was the queen of battle. But large-scale maneuvers by massed armored formations required huge amounts of fuel—more, in fact, than the Third Reich's limited petroleum base could sustain for long without crippling military, naval, and industrial impacts elsewhere. In contrast, the Soviets could draw a more than ample supply from Baku, Grozny, and the other major oil fields of the Caucasus. If they retained that advantage, sooner or later they would become unstoppable. This reality forced the Nazi war plan for 1942. Instead of trying to break Soviet power by resuming the march on Moscow, the Germans would turn their forces south and make an all-out effort to grab the Caucasus.

THE STALINGRAD CAMPAIGN

The most important Caucasus oil field, Baku, is on the Caspian Sea. The heart of the Soviet petroleum industry, it alone was responsible for more fuel production than the entire Nazi empire. The Soviets got their fuel from Baku by using ships to move it along the Caspian shore and then up the Volga River to central Russia. If the Volga transport line could be severed, the Russians would lose their own fuel supply even before the Germans took Baku for themselves. Looking at the map, the Nazi leaders saw one place where the Volga jags sharply to the west, making it the closest point at which the Volga could be cut. The fact that the western river bend also included an industrial town of some importance to the Russian war effort made it a doubly tempting target. Accordingly, the main thrust of the German offensive was directed at Stalingrad.

If the Germans could take Stalingrad, they would effectively cut off not only the Caucasus oil fields but also the Russian forces defending them from the rest of the Soviet Union. Then, with the city serving as a strong point to defend their northern flank, the Wehrmacht's Caucasus invasion forces could overrun the oil-rich region at their leisure.

Thus Stalingrad became the decisive battle of the eastern front, and, arguably, the entire Second World War. The Russians were as keenly aware of its importance as the Germans. The Red Army slogans at Stalingrad were "Not a step back" and "There is no land behind the Volga"—and they meant it. For the Soviets, Stalingrad was do or die. Defending the ruins to the last, they fought back with a tenacity, courage, and sheer ferocity unparalleled in modern warfare. Even when ammunition ran out, they would not yield, and they counterattacked at every turn. In vicious hand-to-hand fighting, the Central Railroad Station changed hands no fewer than seventeen times. Every other strong point in the city—the Tractor Factory, the Univermag Department Store, the Mamayev Kurgan hilltop, the grain elevator district—was fought for with equal abandon.

Against three Soviet divisions, the Germans poured an entire army group of more than a quarter of a million men, and with this preponderance of force gradually seized most of the town. But on November 19, 1942, just when it seemed like the Germans finally had achieved their objective, the Soviets unleashed a massive armored counteroffensive outside the city, and piercing the Axis lines on either side, came together in a pincer movement to surround the German army within.[8]

The German general staff wanted to immediately order their Stalingrad army group (known as Army Group B, commanded by General Friedrich Paulus) to break out of the encirclement and retreat to the west, but Hitler would have none of it. Instead, he ordered a Panzer force to be assembled under one of his best generals, Field Marshall Ernst von Manstein, and launched into the teeth of the Russian armor concentration to break through to the city and support Paulus. There was absolutely no alternative. As Hitler explained to von Manstein, "Unless we get the Baku oil, the war is lost."[9]

Rounding up every reserve Panzer division he could find, von Manstein began his counterattack on December 12. A huge tank battle followed. But even as he slugged it out with the Russian forces blocking his way to Stalingrad, von Manstein became aware that additional Soviet tanks were advancing around his flanks. By December 24 it was clear that if he did not withdraw, his own Panzer group would be surrounded as well.

With the Russians now exploiting their own mobility, von Manstein had no choice but to retreat, and with that decision the thirty-two divisions of German, Hungarian, Romanian, and Italian forces trapped in the Stalingrad pocket were doomed. They held out for another six weeks until, reduced by starvation to cannibalism, they surrendered on February 2, 1943. In the meantime, the Russians seized the initiative to launch their armor in a drive from the Stalingrad front toward Rostov, which borders the Sea of Azov. This threatened to cut off the entire army group (Army Group A) that the Germans had sent into the Caucasus to seize the oil fields. This force had entered the Caucasus during the summer, taken the secondary oil center of Maikop, and slogging on through the mountains, was barely forty miles from Grozny. But now all those hard-won gains would have to be given up. Abandoning most of their heavy equipment, the men of Army Group A rushed to escape from the Caucasus before the Rostov gap closed and they were trapped as well.

Hitler's all-out effort to take the Caucasus oil fields thus ended in utter failure. To compound his predicament, in the same November that Paulus's army was surrounded in Stalingrad, Erwin Rommel's fabled Afrika Corps (itself crippled by fuel shortages) was badly beaten by the British at El Alamein, and then decisively outflanked by American troops landing well to its rear in French Algeria. These developments ended any hope the Germans might have entertained of reaching the oil fields of the Middle East.

So, instead of being reinforced, the threads that held up the fuel position of the Third Reich would be forced to sustain ever-greater loads. With their own Caucasus fuel supplies secure, the growing Russian tank armies would launch one pummeling blow after another, forcing the Germans into a defensive war of maneuver and attrition that they could not afford. Meanwhile, on the North Atlantic in 1943, the Americans would exploit their unlimited fuel supplies to institute nonstop air patrols over the entire ocean route from the East Coast of the United States to Britain, thereby crushing the Nazi U-boat fleet.[10] With the transatlantic route secured, the Americans began to build up a huge army in England with the obvious intent of launching an invasion of the Continent.

With a two-front war in the offing, the strain on the German fuel supply became critical. By the spring of 1944, matters reached the point where one stab in the right place could cut the threads and send the whole Nazi empire into total collapse.

THE BOMBER WAR

In retrospect, it seems odd that the Allies did not realize much earlier the extreme vulnerability of Nazi Germany to a cutoff of its fuel supply. Actually, there were some important Allied military leaders who did, notably General Carl Spaatz of the US Army Air Corps, who became the foremost advocate of what would be known as the Oil Plan.[11] But Spaatz was up against the British, who believed that air power could be used to greatest effect by breaking enemy morale through assaults on cities, as well as other American officers who saw greater merit in hitting German aircraft factories, ball bearing plants, hydroelectric dams, transport centers, or other targets. With the direct support of President Franklin D. Roosevelt, however, Spaatz secured approval to launch one very daring raid on Ploesti in 1943.

So, on August 1, 1943, the US Army Air Corps launched 177 B-24 Liberator bombers from airfields in Benghazi, North Africa, to hit the Romanian oil refineries. Because the round-trip distance to the target was more than two thousand miles, no fighter escort was possible, and the bombers came in alone, at treetop level. Waiting for them were over two hundred scrambled German fighters and a network of hundreds of defensive positions equipped with 88-millimeter antiaircraft guns, all manned and ready.

Seeing this reception committee, the raid's commander, Brigadier General Uzal Ent, is reported to have said, "If nobody comes back, the results will be worth the cost."

Assailed by the swarming fighters, the Liberators, flying at altitudes as low as *thirty feet*, literally dodged among the refinery smokestacks to deliver their loads, while taking fire from flak guns firing

Figure 12.1. USAAF B-24 Liberators hit the Nazi oil refineries at Ploesti, August 1, 1943. Photo courtesy of United States Air Force.

down on them from the surrounding hillsides. As the oil field tanks exploded, more planes were lost flying through the flames. The havoc on the ground was incredible. In less than half an hour, 40 percent of Ploesti's capacity was destroyed. Only 89 of the Liberators that reached the target ever made it home.[12]

The heavy losses experienced at Ploesti deterred the US Army Air Corps from trying again—for a while. But by the spring of 1944, the Americans had the P-51 Mustang, a fighter equipped with drop tanks that gave it the range needed to protect Allied bombers striking targets deep within Germany. With this capability in hand, Spaatz set his sights on Leuna.

At last, the general got his wish. On May 12, 1944, the US Army Air Corps struck the Farben synthetic fuel plants with a devastating 935-bomber attack. With that one raid, the threads finally began to snap. It was a deathblow to the Reich.

In his memoir, *Inside the Third Reich*, Nazi minister of armaments and industry Albert Speer provides a compelling inside view of the collapse of Hitler's empire following the May 12, 1944, raid. Here is what he says:

> I shall never forget the date May 12. . . . On that day the technological war was decided. Until then we had managed to produce approximately as many weapons as the armed forces needed. . . . But with the attack of nine hundred and thirty five daylight bombers of the American Eighth Air Force upon several fuel plants in central and eastern Germany, a new era in the air war began. It meant the end of German armaments production.
>
> The next day, along with technicians of the bombed Leuna Works, we groped our way through a tangle of broken and twisted pipe systems. The chemical plants had proved to be extremely sensitive to bombing; even optimistic forecasts could not envisage production being resumed for weeks. . . .
>
> After I had taken measure of the consequences of the attack, I flew to Obersalzberg, where Hitler received me in the presence of General Keitel. I described the situation in these words: "The enemy has struck us at one of our weakest points. If they persist at it this time, we will soon no longer have any fuel production worth mentioning. Our one hope is that the other side has an air force General Staff as scatterbrained as ours."[13]

Speer did not get his wish. The Americans kept at it. On May 28 they hit Leuna again, and on the following day they blasted Ploesti to pieces. More raids followed. Again, Speer: "On June 22, nine-tenths of the production of airplane fuel was knocked out; only six hundred and thirty two metric tons were produced daily. . . . On July 21, . . . we were down to one hundred and twenty tons daily production—virtually done for. Ninety-eight percent of our aircraft fuel plants were out of operation."

The consequences of the fuel cutoff were felt quickly. Within months, the situation was a total debacle. The statistics are remarkable. In 1944 Nazi Germany actually produced 39,807 military aircraft and 22,100 tanks.[14] But they were nearly all useless due to lack of fuel.

Speer says: "In July, I had written to Hitler that by September all tactical movements would necessarily come to a standstill for lack of fuel. Now this prediction was being confirmed." He goes on to describe how the Luftwaffe was virtually grounded for lack of fuel, and even training new pilots had become impossible because there was no fuel for flight practice.

Speer continues:

> Meanwhile, the army, too, had become virtually immobile because of the fuel shortage. At the end of October, I reported to Hitler after a night journey to the Tenth Army south of the Po. There I encountered a column of a hundred and fifty trucks, each of which had four oxen hitched to it. . . . Early in December, I expressed concern that "the training of tank drivers leaves much to be desired" because they "have no fuel for practicing." General Jodl, of course, knew even better than I how great the emergency was. In order to free seventeen and an half thousand tons of fuel—formerly the production of two and a half days—for the Ardennes offensive, he had begun withholding fuel from other army groups on November 10, 1944.

HITLER'S LAST GAMBLE

The Ardennes offensive—known to Americans as the Battle of the Bulge—was Hitler's last throw of the dice. Every drop of fuel left in the Reich was scavenged to give the remaining Nazi Panzer divisions a final chance to engage in blitzkrieg mobile warfare. According to the plan, they would depend upon overcast December skies to negate

Allied air superiority, and, with the advantage of surprise, break through the weak American infantry forces guarding the Ardennes forest to make a run for Antwerp. Antwerp was serving as the primary port of entry supporting the vast logistics demands of the American and British armies on the western front. If the Germans could take it, the Allied supply position would become impossible.

The plan was clever—almost brilliant in a diabolical way—but it had a deadly weakness: The Germans did not have enough fuel to reach Antwerp. To make the plan work, they would have to capture American fuel supplies. In particular there was one huge depot, near Stavelot in eastern Belgium, that contained 2.5 million gallons of fuel, which they absolutely needed to get. Because of its critical nature, the task of seizing the Stavelot depot was assigned to the most fanatically committed elite unit in the Nazi armed forces, the 1st SS Panzer Division—the Leibstandarte Adolf Hitler—with Heinrich Himmler's ruthless protégé, Obersturmbannführer Jochem Peiper, placed in command.[15]

The Germans opened their offensive on December 16, 1944. Achieving complete tactical surprise, they overran the American positions, taking many prisoners. On the first day, Peiper's rapidly advancing division caught fifty GIs on the roadside near Bullingen manning jerry cans. The Obersturmbannführer made them fill his tanks, after which he had them all lined up and methodically dispatched with pistol shots to the back of their heads. The Nazis then moved rapidly to the west, taking several hundred more Americans captive in the area south of Malmedy on December 17. Not wishing to be slowed down by the need to manage POWs, Peiper ordered the captured GIs herded to an open field, where they were machine-gunned. Peiper gave his men time to crush the survivors' heads with their rifle butts, and then drove on. Storming the bridge across the Amblève River, Peiper led his battle group into Stavelot. After several hours of fierce street fighting, the miscellaneous assortment of American units in the village were overwhelmed and forced to retreat east, leaving the 1st SS free to advance on the giant fuel depot just north of the town.

The only forces available to defend the fuel dump against Peiper's full SS Panzer division were a few lightly armed infantry units who had escaped the debacle at the front, plus some noncombat elements from the 291st and 202nd engineering battalions. With just three puny antitank guns to shoot back, the GIs made the oil their weapon. Frantically, they rolled oil drums into a ditch that ran alongside the dump and set them ablaze, creating an inferno of smoke and fire across Peiper's line of advance. Rather than risk running his Panzers through the flames, Peiper decided to turn west and circle around to try to take the dump from another side.

But as he attempted to flank the fuel depot, the engineers raced ahead of him in jeeps, blowing up one bridge after another before he could cross, until his fuel ran dry. Emerging from the hatch of his immobile Tiger tank, the enraged Peiper reportedly raised his fist at the GIs on other side of the river and railed, "Those damn engineers!"[16]

The fuel depot was saved. Within a week, the whole offensive was on empty. Then the cloud cover broke, and American P-47 fighter-bombers came screaming down out of the sky to blast the paralyzed Panzer divisions. Abandoning their useless vehicles, the Germans began a long hopeless trudge home through the snow.

Nazi Germany had run out of gas.

JAPAN'S WAR FOR OIL

In contrast with the action in the European theater, the Pacific war was not just massively impacted by oil. It was caused by it.

In the 1930s Japan had embarked upon a program of brutal conquest, first of Manchuria in 1931, and then of China, starting in 1937. Shocked by horrifying atrocities committed by Japanese forces during their sack of the Chinese capital of Nanking, the United States attempted to help the Chinese by providing arms to Chiang Kai-shek's nationalists, and also by implementing a series of escalating trade sanctions against Japan. Despite this, the Japanese were successful in overrunning most of the country within a few years.

The growing Japanese empire faced a major strategic difficulty, however. It had virtually no petroleum resources. The closest major oil fields were those of the Dutch East Indies, in present-day Indonesia. What if foreign shipments were cut off? In the view of the imperial cabal, the empire would never be secure until the Indonesian oil fields were taken.

The Nazi conquest of western Europe in 1940 gave the militarists their chance. With their homelands overrun, the French and Dutch colonies in Southeast Asia were virtually defenseless. Taking advantage of the situation, in the summer of 1941 the Japanese seized the French colony of Indochina, thereby positioning themselves within striking range of the Dutch East Indies. The United States responded by declaring a total oil embargo against Japan.

The US move was intended to put pressure on the Japanese to deter them from further conquests, and, if possible, induce them to withdraw from Indochina. Instead it had the effect of convincing the Japanese leadership that the United States would intervene militarily if they took their next step and invaded the oil-rich Dutch islands as planned. The Americans ruled the Philippines and had strong military bases there, virtually astride the maritime routes that would connect Japan to its planned Indonesian conquests. Operating out of these bases, hostile American air and naval forces would be able to cut off any tanker deliveries of oil from the East Indies to Japan. This threat could not be tolerated. Therefore, the Philippines would have to be conquered as well. This meant war with the United States. Given the US advantages in population and industry, victory could be achieved only by a sudden strike that would disarm the Americans with a single blow. Thus the Japanese laid their plans for a surprise attack on the US Pacific Fleet docked at Pearl Harbor.

The decision by the Japanese to initiate war with the United States was insane, as they were picking a fight with a power that was completely out of their league. To cite just one example, during the entire war, Japan produced 589 warships of all kinds, versus 8,812 (not counting landing craft and auxiliary vessels) produced by the United

States. Measured in weight of war machinery, the Ford Motor Company *alone* outproduced the entire nation of Japan. The United States was technologically superior as well. And while the magnitude of these disparities may not have been apparent before the war, the basic fact of their existence was clear enough. Had the Japanese not attacked Pearl Harbor, but simply moved south to take the Dutch East Indies, there is every possibility that the Roosevelt administration, boxed in by isolationists, might have limited its response to sanctions and denunciations. After all, the Americans had not been willing to fight to save Britain, let alone Holland; was it really so certain that they would declare war to protect some Dutch colonial possessions? Yet Japan's leaders were so frenzied over the possibility that the United States *might* intervene militarily to cut their oil supplies that they forced the issue themselves.

As a result of its irrationality, however, the Japanese onslaught was unexpected and achieved considerable tactical success. In their surprise attack on Pearl Harbor, the aviators of the Imperial Japanese Navy sank or disabled all seven battleships of the US Pacific Fleet, and destroyed hundreds of land-based aircraft parked wingtip-to-wingtip on Hickham and Wheeler fields. Ironically, however, despite their keen awareness of their own oil supply vulnerability, the Japanese neglected to target that of the Americans. At Pearl Harbor, not far from battleship row, the US Navy kept its oil supply in large aboveground fuel tanks. Had the Japanese hit them, they would have paralyzed the entire US Pacific Fleet. But they didn't. The consequences of this failure were immense. As US Navy Pacific Fleet commander Admiral Chester Nimitz put it later: "All the oil for the Fleet was in surface tanks at the time of Pearl Harbor. We had about four and a half million barrels of oil out there and all of it was vulnerable to .50 caliber bullets. Had the Japanese destroyed the oil, it would have prolonged the war another two years."[17]

In contrast, the US Navy was quick to zero in on fuel logistics as the key Japanese weakness. American submarines operating in the Pacific were instructed to prioritize oil tankers as their number-two targets, second in importance only to enemy aircraft carriers. The effects of this order were devastating. As merchant raiders, the US subs of the World War II Pacific were far superior to their German U-boat counterparts in the Atlantic, and they faced an opponent whose antisubmarine capabilities were primitive in comparison to the Anglo-American convoy system. In consequence, they were able to take an enormous toll on the Japanese merchant fleet.[18]

Thus, while the Japanese did manage to capture the huge oil fields of the Dutch East Indies in early 1942 and bring them into operation despite considerable wrecking done by retreating Allied operatives, by 1943 they could no longer move the oil effectively from Indonesia to Japan. This created extreme fuel shortages in the Japanese home islands and made it impossible to train new pilots adequately. So, in the battle of the Mariana Islands in June 1944, the Japanese were constrained by fuel shortages to approach the American fleet directly, rather than maneuver strategically, and to send virtually untrained replacement pilots into aerial combat with experienced US aviators. The result was the lopsided engagement known to Americans as the "Great Marianas Turkey Shoot" in which the Japanese lost 273 planes. The Americans lost 29.

With the oil tanker route from the Dutch East Indies to Japan largely cut, the Japanese were left with a choice of whether to base their navy in Japan, where their ships could get no oil, or in the Indies, where they could get no ammunition. In the end, they decided to base their battleship fleet in the Indies and their carriers in Japan. It was with their fleets thus uncomfortably divided that they were forced to try to find some way to resist the American invasion of the Philippines, which began at the island of Leyte in October 1944.

The Japanese had gone to war with the United States in order to secure their oil supply route by seizing the Philippines. Now they would have to risk everything to try to hold them. As Admiral Soemu Toyoda, chief of the naval general staff, later put the matter: "Should we lose in the Philippines operations, even though the fleet should be left, the shipping lane to the south would be completely cut off so that the fleet, if it should come back to Japanese waters, could not obtain its fuel supply. If it should remain in southern waters, it could not receive supplies of ammunition and arms. There would be no sense saving the fleet at the expense of the loss of the Philippines."[19]

So with their backs once again to the wall over the issue of oil logistics, the Japanese launched a desperate plan. Sailing down from Japan with the empire's last remaining carriers, Admiral Jisaburo Ozawa would launch an air strike upon the nine fleet aircraft carriers, eight light carriers, six modern fast battleships, seventeen cruisers, and sixty-three destroyers of Admiral Bill Halsey's Third Fleet (aka Task Force 34). The primary purpose of this attack would not be to inflict damage, which was very unlikely with the weak air forces at Ozawa's disposal, but rather to draw the Third Fleet away to the north in pursuit. Another task force consisting of two obsolete battleships and some cruisers and destroyers under Admiral Shoji Nishimura would enter the Surigao Strait to the south of Leyte to draw away and engage the six older American battleships covering the landings. Thus Japan would sacrifice its carriers and its southern task force as "decoy ducks" to allow Admiral Takeo Kurita's fleet of five battleships and twelve heavy cruisers to sail though the San Bernardino Strait to the north of Leyte and strike the American troopships carrying the hundreds of thousands of GIs of Douglas MacArthur's invasion force. Japan would lose most of its navy, but the US Army would be slaughtered.

The plan nearly worked. As at Pearl Harbor, the incredible nature of the Japanese strategy protected it from American comprehension until it was too late. Taking the bait to the north and the south, the primary combat elements of the American fleet sailed off, leaving the troopships unprotected except for an antisubmarine force of twenty-four destroyers and eighteen escort carriers under Admiral Thomas Kinkaid. Emerging from the San Bernardino Strait at dawn of October 25, Kurita's powerful fleet, led by the super-battleship *Yamato*, took Kinkaid's light task force by surprise.

What followed was one of the most desperate battles in the history of naval warfare. With a general massacre of the GIs packed helplessly in their troopships imminent, the little "tin can" destroyers of Kinkaid's squadron hurled themselves furiously against the giant battlewagons of Kurita's fleet. While shells burst the decks of their escort carriers, antisubmarine patrol aircraft armed with depth charges took off through the fire to try to counterattack. Braving the massive shells of the enemy battle line, Kinkaid's destroyers closed the range and launched a salvo of torpedoes that sent Kurita's fleet turning every which way in confusion. Swerving away to avoid being hit, Kurita's *Yamato* found herself bracketed by two torpedoes, and, despite her eighteen-inch guns, was forced to steam away from the battle for twenty minutes until they ran their course. The battle became a chaotic free-for-all as the tiny American destroyers, their five-inch pop guns blazing, charged about launching torpedoes amid a fleet of crazily maneuvering Japanese warships ten times their size.

But as brave as they were, the Yankee tin cans were outgunned in weight of firepower by more than a hundred to one, and it seemed like the battle could only end one way. Then suddenly, at 9:20 in the morning, with the annihilation of Kinkaid's squadron and the nearby troopships virtually within his grasp, Kurita ordered a general withdrawal. As if by a miracle, MacArthur's army was saved, and with the loss of the two decoy fleets, the Battle of Leyte Gulf became an unqualified disaster for the Japanese.

Kurita's retreat seems inexplicable. Japan had sacrificed its last aircraft carriers and many other ships to place him in a position to deal a devastating blow to America. Yet at the brink of triumph, he turned away. Why? Military historians have debated the issue ever since. Some have said that the fury of the destroyers' onslaught simply unnerved him, or bluffed him into believing that the arrival of powerful American reinforcements was imminent. The Japanese naval historian Masanori Ito, however, offers a different explanation. According to Ito, who knew many of those present on the *Yamato*'s bridge when the decision was made, Kurita had no choice but to withdraw because the three hours of unplanned high-speed maneuvers in the chaotic engagement with Kinkaid's destroyers had dangerously depleted his fuel supply.[20]

At Leyte Gulf, Kurita had a hundredfold advantage in firepower and armor. But without oil, he couldn't win.

Its navy all but gone, Japan was helpless against the American advance. Their garrisons isolated, one island fell after another. Then, on August 6, 1945, a single American bomber arrived over the city of Hiroshima. Despite the fact that Japan had produced more than eleven thousand aircraft during 1945, the *Enola Gay* needed no fighter escort, for none of the Japanese planes had any fuel with which to fly. One bomb was dropped, and in an atomic flash, the city was destroyed. After a second attack a few days later, Japan surrendered.

THE SCALES OF VICTORY

It would be false to argue that World War II was won by the Allies' superiority in oil. Wars are won not by oil or other resources, but by the courage and self-sacrifice of the people who do the fighting. Yet, without question, it was the control of oil that tipped the scales of victory. To see this, just consider how the war would have gone if, instead of being located in the United States and Soviet Union during the 1940s, the primary developed oil resources of the world had been inside Germany and Japan.

Had that been the case, Britain would have fallen in 1940, as all the Nazis would have needed to do to ground the Royal Air Force and immobilize the Royal Navy would have been to declare an embargo on oil exports to the United Kingdom. Similarly, deprived of oil, the Soviets would never have been able to create mobile tank forces, and their static infantry armies would have been repeatedly surrounded and ultimately annihilated by German Panzers on the open steppe. With a secure supply of oil, Japan would never have had to attack the United States, and so could have consolidated her Asia-wide empire at leisure, especially since there would have been no way for an oil-deprived America to initiate any offensive action to interfere. The entire Eurasian supercontinent would thus have been secured for the Axis, and, that done, the rest of the world would soon have followed.

This brings us to the present day.

America currently faces a foe whose power rests almost exclusively on their control over oil. It is the revenue from oil that is allowing Saudi Arabia to finance the global propagation of the Islamofascist movement as well as the systematic corruption of many Western governments, including our own. It is revenue from oil that is providing Iran with the wherewithal to develop the atomic bombs that will give its Hezbollah terrorists—who are now expanding their operations to the Western Hemisphere—the capacity to slaughter millions of people. It is our dependence upon the oil controlled by such enemy powers that is preventing us from undertaking effective action against them. It is their control over oil vital to us that allows the Islamists to laugh in the face of our complaints as they teach terrorism, sharpen their nuclear knives, and call for our doom.

In World War II, we controlled the oil. In this war, the enemy does. This is an unacceptable situation, because it places our fate in the hands of people who want to kill us. In World War II, we had no compunction about destroying the Nazi fuel-making facilities at Ploesti and Leuna, or of systematically sinking the Japanese tanker fleet—*because we didn't need their oil.* As we have shown, those attacks were incredibly effective in breaking enemy power. On May 12, 1944, the day of the Leuna raid, the Third Reich ruled an empire comprising nearly all of continental Europe, with a collective population and industrial potential exceeding that of the United States. A year later, it did not exist. Once its tanker fleet was sunk, the collapse of the Japanese empire was almost as fast. Today we are confronted by an enemy without a shadow of the armaments of the Axis; all they have is oil. Were we to destroy that power, they would be left with nothing at all. But we can't hit them where it would truly hurt, because our economy needs their oil to survive.

Consider the current crisis with Iran. The fanatical Iranian government is developing nuclear weapons. How can we stop them? The nuclear facilities themselves are scattered and well dug in, so effectively striking them may be difficult. In the face of such a problem, the foreign policy leaders of the Bush administration have been reduced to impotent whining about the Iranian program in the UN. However, the entire nuclear program could be rapidly shut down—along with the rest of the Iranian government—by cutting off the oil income that is paying for it. This could readily be accomplished by launching a modest air strike on Iran's oil export terminal on Kharg Island. Since this facility is replete with large thin-walled tanks filled with very flammable petroleum, the delivery of a dozen precision-guided bombs would probably suffice to do the job. The only thing barring us from such a course of action is the consequence to the global economy of the loss of Iran's oil. However, if a situation could be brought about where the world no longer needed Iranian petroleum, the task of removing it from market would present no difficulties whatsoever.

The Saudi oil export system also has similar points of complete vulnerability. For example, there is the giant oil terminal at Ras Tanura, which handles some 80 percent of the Wahabbi kingdom's exports. Another delicate spot is the oil-refining complex at Abqaiq, which feeds Ras Tanura about 7 million barrels per day. Some ungrateful lunatics from al Qaeda actually attempted to take the place out with a car bomb in February 2006, so it's probably going to be

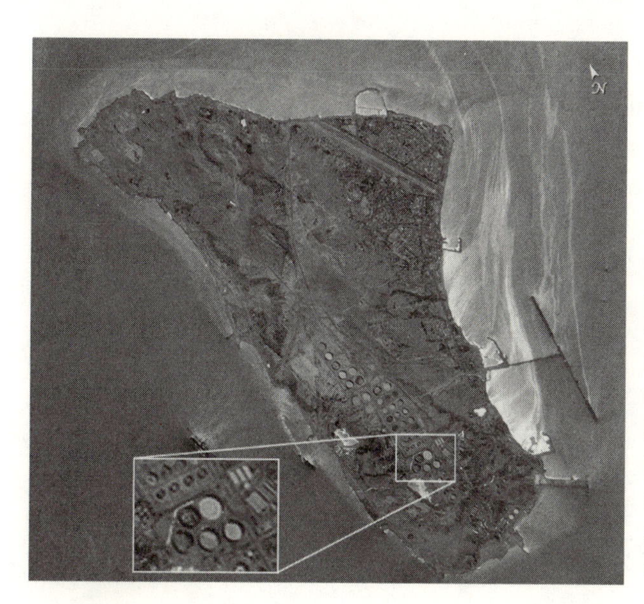

Figure 12.2. Located at 29° 22' N, 50° 30' E, the Kharg Island terminal handles 80 percent of Iran's oil exports. Note the highly vulnerable petroleum storage tanks. Photo courtesy of NASA.

destroyed eventually no matter what we do (which is yet another reason to make it unnecessary as soon as possible). However, that said, were it in our interests to do so, we could shut the place down in an afternoon.

So the crux of the matter comes down to this: Do we want to win or lose? The issue at stake in energy security is not a matter of whether the price of gasoline will be $2 per gallon or $3 per gallon; it is who will determine the human future. Do we want to have the enemy's fate in our hands, or do we want to have ours in theirs?

CHAPTER 13
ENERGY AND STATECRAFT

For the past three decades, US energy policy has been a scandal. In the face of repeated petroleum blackmail, the response of the political class has been not only to fail to take effective action and make us energy secure, but also to allow our petroleum dependence to *double*, from 30 percent to 60 percent. As a result, we now face a severe national security crisis and a potentially fatal weakness. We need to eliminate that weakness.

In his 2006 State of the Union address, President Bush made a ringing declaration that America must free itself of its addiction to foreign oil. Yet his accompanying policy initiatives included nothing that would accomplish that objective. Instead, the administration simply asked for $150 million in research money for cellulosic ethanol and set a rhetorical target of reducing oil imports by 20 percent by 2025.

Following the State of the Union address, I got a firsthand taste of the administration's fecklessness on the energy issue when I sent the president's science adviser, Dr. John Marburger, an advance copy of my March 2006 article in the *American Enterprise* magazine, which presented in summary form the flex-fuel program for energy independence laid out in this book. Marburger read the article closely and asked me a number of detailed questions, which I answered. I then asked him, "OK, so why not implement the plan? If the president introduced a bill calling for a flex-fuel mandate, he'd get bipartisan support and the bill would pass. It would be a real accomplishment for the administration and for American energy independence." Marburger answered: "We don't believe in mandates."

I subsequently met for two hours with one of Marburger's senior staff. While finding the idea of moving to flex-fuel cars interesting, he objected to the concept in that it would cost the American auto industry a total of $150 million to make the necessary conversion. This is less than the United States spends on foreign oil every *five hours*.

Congressional action has also been inadequate. In the fall of 2005 a group called the Set America Free Coalition (SAFC, of which more below) circulated the draft of a bill that contained a number of elements helpful to advancing the cause of energy independence. Senator Evan Bayh (D-IN) and Congressman Jack Kingston (R-GA) introduced it into their respective chambers, and upon being lobbied by the SAFC, twenty-seven senators and eighty-seven congressmen subsequently signed on as cosponsors. On the Senate side, however, staff watered the bill (S.2025) down so that it contained no effective provisions to move the nation toward flex-fuel cars, making it essentially useless. As a result of this failure, three senators (Tom Harkin [D-IA], Richard Lugar [R-IN], and Barack Obama [D-IL]) introduced a separate bill (S.1994) that would mandate that all new cars be flex-fueled by 2015. However, with the exception of Maria Cantwell (D-WA), no other senators agreed to cosponsor, the rest choosing to stick with the fake S.2025 bill instead.

The House version of the SAFC bill, HR.4409, emerged much better. Defining a flex-fuel car as one capable of burning any combination of gasoline, E85, or M85, the bill mandated that 80 percent of all new cars *manufactured* in the United States be either flex-fuel or biodiesel-powered by 2011, and 90 percent by 2012. That would certainly be a real step forward. In addition to the unnecessarily extended implementation time line, however, the bill included a serious deficiency in that it imposed no requirement that foreign cars *sold* in America be flex-fueled. This omission is very bad because mandating that foreign cars sold here be flex-fueled is the key for forcing the transition to the alcohol economy worldwide, and if we actually want to break OPEC, nothing less will do.

Unfortunately, while getting eighty-seven House cosponsors for such a bill was a major accomplishment for the SAFC, in face of counterlobbying by the auto industry and the OPEC fifth column (and in the absence of any support from either the White House or Speaker of the House Dennis Hastert [R-IL]), the bill ultimately went nowhere.

Still, every year is always a new ball game in Congress, and in 2007 the SAFC is trying again with a new bill called the DRIVE Act, which does include a flex-fuel mandate.[1] This organization, based in Washington, DC, has assembled in its roster a nonpartisan array including national security hardliners Frank Gaffney and Clifford May, former CIA director James Woolsey, former Senate minority leader Tom Daschle, and Nobel Prize–winning chemist George Olah, among others, and me. Its real leaders, however, are Gal Luft and Anne Korin, two remarkable people who have made the achievement of American energy independence their passion and crusade. If you care about the future of this country, I urge you to support their efforts. You can get involved by reaching them through their Web site at setamericafree.org.

Again, we face a crisis. The enemy is using our oil payments to foment a global nuclear-armed jihad, and we can't lay a finger on them because we depend on their oil. As Luft summarized it in an October 6, 2006, op-ed in the *Baltimore Sun*:

> Winning the war on radical Islam at current oil prices is an oxymoron. Unlike World War II and the Cold War, in which America's strategy was to deny the other side the economic wherewithal necessary to keep up the fight, in the current war, the U.S. is doing the exact opposite. While the U.S. economy hemorrhages nearly $300 billion a year to pay for imported crude, oil-producing nations such as Saudi Arabia and Iran that are sympathetic to—and directly supportive of—radical Islam are on the receiving end of staggering

> windfalls. . . . An undetermined portion of that money finds its way—through official and unofficial government handouts, charities and well-connected businesses—to the jihadists committed to America's destruction.
>
> In order for us to regain control over our foreign policy, reduce our vulnerability to supply disruptions and curb the flow of petrodollars to unfriendly regimes, a new and central objective should be added to our foreign policy: leading the world on an accelerated shift, enabled by modern technology, toward a global transportation system based on next-generation, nonpetroleum fuels and the cars and trucks that can run on them.

The method for doing so is clear. Congress needs to pass a law mandating that all new cars sold in America be flex-fueled. This will force all automobiles produced worldwide to be made flex-fueled, thereby creating a huge global market for alcohol fuels. Some of these fuels, such as methanol, can already be produced substantially more cheaply than gasoline, and only need the market provided by a flex-fuel mandate to drive enormous production capabilities into existence. Others, such as ethanol, are potentially vulnerable to OPEC pricing maneuvers and may need tariff protection to prevent periodic Saudi dumping from derailing them. When necessary, such protection should be provided. We should not sell our freedom for a mess of potage.

If we adopt such a deliberate goal-driven policy to grow the alcohol economy, we can make OPEC's oil unnecessary. We will then be in a position to dictate the terms of capitulation to the terror bankers. Instead of being their shield, the enemy's petroleum will become their greatest vulnerability. Under such conditions, it will probably prove superfluous to actually strike the Saudi oil facilities. In a game of chess, the struggle ends not with the taking of the enemy king, but with his entrapment. Once we are energy independent, the enemy will be rendered helpless, and, one way or another, their oil for terror game will have to fold.

Call it checkmate. Call it victory.

OIL AND FREEDOM

The enemy's control of the global oil supply is a threat to the freedom of all people around the world. Indeed, the worst victims are the populations of the petrotyrannies themselves, for by giving their governments unlimited expense accounts, it endows the powerful with the ability to trample their subjects at will.

Gifts make slaves. In European history, the acquisition of vast unearned New World treasure by Spain led to the degradation of the enterprising classes and caused the abortion of the development of representative institutions in that country. In contrast, the need of England to develop its sources of revenue through commerce and manufacturing forced the British monarchy to consult with Parliament and enact measures suitable for growing the productive powers of the nation. Similarly, in the Middle East today, those nations with the least oil have the most literacy, education, and freedom. The people of Arabia will never be free so long as their rulers have no need of their human capacities.

The enemy control of oil also threatens our own freedom, in three different ways. On the one hand, there are those who would use the need to counter the oil cartel as a pretext to demand measures of economic strangulation, thereby furthering their quest to expand state power at the expense of the individual. On another, there are those who would respond to the perils posed by the terrorists funded by the enemy oil with ever more intrusive police actions. The Bush administration has already begun to enact measures along such lines, and more will certainly follow, regardless of who is in power, if the terror attacks are allowed to continue. It is an axiom of political science that liberty will not thrive in a state of siege.

But worse even than these threats from the Left and the Right is the menace posed to freedom by the oil-empowered enemies themselves. What happens if they win?

WHY WE FIGHT

It is impossible to win a war unless you know what you are fighting for. It is impossible to win a war if you do not know who and what the enemy is. This war may be decided by control over oil, but it's not about oil. It's about ideas—very portentous ideas. If we are going to win this war, we need to understand what these ideas are, and where they came from. So before we bring this book to a close, let us reexamine these most central issues, starting at the very beginning.

The Jewish creation story recorded in the book of Genesis bears many resemblances to the contemporary Babylonian myth of origins as told in their Enuma Elish. Thus, for example, in Genesis, God begins the world by separating "the waters above from the waters below," while in the Enuma Elish, the chief god Marduk gets things going by cutting the sea goddess Tiamut into upper and lower halves. Such parallel views are not surprising, as the Bible itself reports that the Hebrews originated in Babylonia. Indeed, in their respective tales, both God and Marduk go on to make men out of clay. But then the accounts differ sharply. According to the Jewish version, God created man "in his own image," breathed his own spirit into him, blessed him, and presented him with dominion over nature. In the Babylonian version, Marduk made men to be his *slaves*. Furthermore, in the Marduk cult as it developed in Babylonia, and under other names such as Baal elsewhere, the enslaved position of man before Marduk was to be acknowledged by the regular ritualized self-abasement of human sacrifice, meaning not merely the murder of captives (which was frequently done by most ancient peoples), but the killing of firstborn children by their parents as a terrified offering of appeasement. In contrast, the Hebrew Bible pointedly rejected human sacrifice, both through the anecdote of Abraham and Isaac told for that purpose as well through legislation laid down later by Moses in the book of Exodus.

Christianity took the Jewish rejection of human sacrifice a dramatic step further by contending that not only did God not demand that man die for him, but God was willing to die for man. Furthermore, rather than viewing God as a cruel superbeing transcendently above his human slaves, the Christians chose to view him as "Our Father," which, taken reciprocally, implies that humans are children of God.[2] To these forcefully humanist theological assertions, the Western heritage added the radical individualist proposition advanced by the Greek philosophers Socrates and Plato that there is an innate faculty of the human mind capable of distinguishing right from wrong, justice from injustice, truth from untruth. Embraced by early Christianity, this idea became the basis of the concept of the *conscience*, which thereupon became the axiomatic foundation of Western morality. It is also the basis of our highest notions of law—the Natural Law determinable as justice by human conscience and reason, put forth, for example, in the US Declaration of Independence ("We hold these truths to be self evident . . .")—from which we draw our belief in the fundamental rights of man existing independently of any laws that may or may not be on the books or existing accepted customs. It is also the basis for *science*, man's search for universal truth through the tools of reason.

As the great Renaissance scientist Johannes Kepler, the discoverer of the laws of planetary motion, put it: "Geometry is one and eternal, a reflection out of the mind of God. That mankind shares in it is one reason to call man the image of God." In other words, the human mind, because it is the image of God, is able to understand the laws of the universe. It was the forceful demonstration of this proposition by Kepler, Galileo, and others that let loose the scientific revolution in the West.

Science, reason, morality based on individual conscience, human rights: This is the Western humanist heritage. Whether expressed in Jewish, Christian, Hellenistic, deist, or purely naturalistic forms, it all drives toward the assertion of the inalienable dignity of man.

Fundamentalist Islam denies all of this. It denies the existence or deserved authority of the conscience. Instead, right and wrong can be known only through the Koran, as interpreted by fundamentalist mullahs. It denies moral responsibility further because it denies the existence of free will. It denies reasoned investigation of nature completely because it denies the idea of causality.[3] Instead, it argues that the universe is created and destroyed repeatedly in every succeeding instant by the will of Allah. Thus scientific activity is useless, and in fact is proscribed.

For Christians, humans are *children of God.* For Islamists, as for the Marduk-Baal cult, they are *slaves of God.* As such, they are also slaves of the master that Allah has chosen to place over them. This view makes democracy and human rights impossible. It also makes social progress impossible, because it denies the legitimacy of any reasoned critique of traditional arrangements. It also makes peaceful coexistence with free non-Muslims impossible, because it denies their right to exist.

As Ibn Taymiyyah, the founding theologian of the Saudi Wahabbi state cult, put it: "Jews and Christians, as well as Zoroastrians, must be fought until they embrace Islam or pay the *jizya* [a special annual ransom paid by non-Muslims for permission to live] without recriminations. . . . When they have not been killed in the war, they must in any case be reduced to slavery. The wives of infidels must also be reduced to slavery and the possessions of infidels must be confiscated."

The infidels *must* be enslaved and their possessions confiscated, Ibn Taymiyyah says, because "Allah has created the things of this world only in order that they may contribute to serving him, since he created man only in order to be ministered to. Consequently, the infidels forfeit their persons and their belongings which they do not use in Allah's service to the faithful believers who serve Allah and unto whom Allah restores what is theirs."[4]

In other words, according to Ibn Taymiyyah, it is the sacred *duty* of Muslims to wage armed jihad against infidels to take away their freedom, wives, children, land, and other possessions, which they are holding illegally in defiance of Allah's will. The infidels are thus not merely legitimate prey, they are rebel outlaws, enemies of Allah who must be hunted down and either killed or enslaved and dispossessed.

This is the fundamentalist Islamic creed. It is to spread this doctrine that the Saudis have set up more than twenty thousand madrassas around the world. As the Saudi government puts it in its official English translation of the Koran: "Fight those who do not believe in Allah nor the Last Day, nor hold forbidden which has been forbidden by Allah and His Messenger, nor acknowledge the Religion of Truth, (even if they are) of the People of the Book, until they pay the Jizya with willing submission, and feel themselves subdued" (sura 9:29).[5]

It was on the basis of this imperative that the fanatical fundamentalist hordes of early Islam fanned out from Arabia to subdue the Christian and Jewish populations of the Byzantine Middle East, North Africa, and Spain, as well as the Zoroastrian people of the Persian Empire. Apologists for Islam make great claims for its "tolerance," because in contrast to conquered polytheists, who were simply massacred, the above-cited people of the book were allowed to continue to exist as dhimmi subjects under Muslim rule. Such claims grossly distort the historical record. In fact, the regime set up by victorious Islam was one of systematic degradation, oppression, and ultimately, extermination of the dhimmis. Dhimmis were made to wear distinctive clothing to mark them out for victimization. As a price for their survival, they were made to pay a yearly ransom. This jizya had to be paid in person in a public place, so the dhimmi making the payment could be held by the beard and beaten for purposes of humiliation at the same time. Dhimmis were forbidden to ride horses or to own weapons, and could not bring suit or testify against Muslims in courts of law.[6] This effectively made them helpless against any crime any Muslim chose to inflict upon them, including theft and rape of their wives and children. Mass expropriation and enslavement of dhimmi children was also practiced as regular policy by Islamic governments from the time of the initial conquests right on down through the twentieth century.[7]

To all this, the Islamic regime added terror in the form of repeated massacres, ranging in scale from the destruction of villages and urban neighborhoods to the slaughter of entire populations. The genocide of 2.5 million Christian Armenians under Ottoman rule between 1894 and 1917 is the instance best known to most Americans, because American missionaries brought it to the attention of the US press at the time.[8] (It was during its coverage of these massacres that the *New York Times* first coined "holocaust," as a new term for mass extermination, which later found use as a means for describing the subsequent Nazi brutalities directed against the Jews.) In fact, however, the early twentieth-century mass killing of Armenians was not exceptional, but was rather typical of numerous comparable slaughters of Christian, Jewish, and Zoroastrian subjects of Islamic occupation going back thirteen hundred years. Early dhimmi victims of Islamic massacres included the Zoroastrian population of Persepolis, Coptic Christians in Egypt, and Catholics in Spain. Large Jewish populations in North Africa and Yemen were also slaughtered. It was the news of such atrocities committed against the Christian inhabitants of Jerusalem, Damascus, Ramlah, and Gaza in the eleventh century that spurred Europeans to launch the First Crusade. The list goes on, and does not stop. In the nineteenth century, in addition to Armenians, Greek, Serbian, Bulgarian, Mesopotamian, and Lebanese Christians were subject to repeated massacres. In the twentieth century, fifty thousand Nestorian Assyrian and Chaldean Christians were exterminated in Iraq, as were more than a million Greeks in Asia Minor, while terror continued against the Copts in Egypt, the Maronites in Lebanon, and other Christians from Biafra to Pakistan, as well as the remaining Jewish populations of North Africa, Yemen, Mesopotamia, and Persia.

As a result of the combined effect of these murderous campaigns, the once-dominant Christian and Zoroastrian population of the Middle East has been nearly eliminated. Such is the program of Islamist tolerance for the people of the book. As for others not so highly regarded, such as African animists and South Asian Hindus, Buddhists, Taoists, and Confucians, no such claim of tolerance is made.

Now it is true that in the Islamic world the fundamentalists have not always been in control. In its formative period, Islamic society included a strong rationalist current led by the Mu'tazilites, who believed in the parity of reason and revelation, and produced many profound philosophers such as Al-Farabi, Averroës (Ibn Rushd), and Avicenna (Ibn Sina). As a result of its good fortune in having conquered the most economically developed parts of the defunct Roman Empire, as well as Persia, the Islamic world was much more urbanized than early medieval Europe and had a higher level of material culture and literacy. A thousand years ago, it was not the West, but Islam, that had the broadest intellectual horizons. Interacting with subject dhimmi intellectuals,[9] Islamic thinkers created algebra and significantly advanced astronomy, optics, and medicine. At a time when there were no colleges in Europe, the Islamic world had hundreds. At a time when the largest European libraries contained a few hundred volumes, there were Islamic libraries with hundreds of thousands.[10]

But as time went on, the ongoing degradation of the dhimmi population removed an important source of cosmopolitan intellectual stimulation within the Islamic world. Then reaction set in, as the fundamentalists denounced the entire enterprise of rational investigation of nature to be intrinsically heretical. The Mu'tazilites were suppressed and the philosophers were made into fugitives. Scientific inquiry was banned. Libraries that were found to contain scientific works were burned. Printing, which appeared briefly in the Islamic world several hundred years before its advent in Europe, was banned, and did not reappear until its reintroduction by American missionaries in the 1830s. The colleges were turned from centers of inquiry into mental slaughterhouses, where generation after generation of the brightest youth were made to rote memorize the Koran.

With the fundamentalist takeover, one of the most developed societies to appear thus far in human history was turned into a wasteland of misery, poverty, mental slavery, and ignorance. A quarter of the world was turned into a graveyard of the mind, which, for the past seven hundred years, has not produced a single significant scientific advance.

In addition to intellectual death, fundamentalist Islam also offers the world the prospect of absolute political tyranny, directed against not only despised dhimmis but Muslims as well. Those who accept their definition as slaves of Allah must also accept their role as slaves of those who Allah has appointed to be their masters.[11] Thus, in Saudi

Arabia, the population is defined to be the *property* of the monarchy, and commoners can be—and are—dispossessed by members of the royal family at will. There can be no such thing as the rule of law under fundamentalist Islam, since neither Allah nor his appointed representatives can be bound in any way. There is only shari'a, a barbaric set of seventh-century Arabian customs, notably involving the degradation of women through encloisterment and polygamy, and therefore the vicious brutalization of the majority of males as well.[12]

THEIR PROGRAM AND OURS

> We hold these truths to be self-evident, that all men are created equal, that they are endowed by their Creator with certain unalienable Rights, that among these are Life, Liberty and the pursuit of Happiness.—That to secure these rights, Governments are instituted among Men, deriving their just powers from the consent of the governed.
>
> —The Declaration of Independence, 1776

As a modern-day incarnation of the cult of Marduk, fundamentalist Islam is by no means unique. We have seen several other versions recently. For example, there are Nazism, fascism, and totalitarian communism. Whether it is for Allah, the Volk, or the proletariat, such movements all amount in the end to the same thing—the sacrifice of the human individual on the altar of the transcendent entity. The resemblance between the forms of oppression of Islamism for the dhimmis and the Third Reich for the untermenschen is particularly striking, and not coincidental, but we must cast our glance more broadly in order to see the phenomenon in its entirety. When former vice president Al Gore says that "[w]e must make the rescue of the environment the central organizing principle of our civilization,"[13] he is coming dangerously close to proclaiming his loyalty to the same camp. When David Pimentel or others of his ilk call for stopping economic growth and halving population, they are waving its banner. While differing in its external costume, the Malthusian view of humans as vermin whose aspirations must be suppressed in order to preserve a fixed order of nature is, at its core, essentially the same as that of Islamofascism—and ultimately can only be enforced by comparable tyrannical means. Both cults demand human sacrifice. Both deny human dignity. Neither can tolerate freedom.

The founders of our country did not believe that the purpose of government is to serve Allah or the environment, but to ensure liberty. We need an energy policy that will do so.

We are at war not with a few thousand cultists, but with a cult. To defeat the enemy, we must not only destroy its current forces, we must discredit the ideology that allows it to recruit. We are not at war with an assortment of savages. We are at war with a set of ideas that reduces people to savages.

The first step required for victory is to disarm the forces that are actively attacking us. This can be done most effectively by taking away their money. As the primary financiers of the global promotion of Islamofascism, the Saudis absolutely must be broken. If that is not done, the fire will continue to spread out of control. As the second most important backers of Islamist terrorism and a state openly committed to developing nuclear weapons for the purpose of raising terrorism's capacity for genocide, Iran must also be denied the funds necessary to achieve its purpose. Both of these objectives can be achieved by instituting an energy policy that makes the oil exports offered by these enemies expendable.

But we need to do more. To defeat fundamentalism and other forms of reactionary totalitarianism, we need to show the power and nobility of human reason, on every level, from the most practical to the most spiritual. How can this be done?

The key is world development. This is important not just for economic and humanitarian reasons, but to show the power of reason in action. The enemy's ideology is stasis; ours is progress. We can prove its merit. Parents who see their children saved from disease by modern medicine can learn that the miracle they are witnessing is not the result of magic, but of scientific thinking. Higher living standards will make mass education and literacy possible. Rural electrification will not only bring economic progress and freedom from drudgery, but enable the widespread proliferation of radios, televisions, and computers, thus opening up previously closed societies to cosmopolitan ideas.

The alcohol economy is the way to achieve both of these objectives. By making a willful strategic decision to switch from petroleum to alcohol fuels, we can redirect trillions of dollars that are now going to the terror bankers to finance death, and send it to the world's farmers to create life. Instead of continuing to accept the hideous injustice of global poverty, to our eternal shame, we can take the funds we are now giving to the enemy and use them instead to launch a new age of world development. We can protect our own freedom by ending the terrorist siege that is provoking responses that endanger it, and by refuting through the fruits of our ingenuity the arguments of those who claim we need tighter social controls to enforce our acceptance of stasis. We can spread liberty by creating a world economy that transfers power from those who *take* their wealth, pre-made, from the ground (and therefore have no need for education or freedom), to those who *make* their wealth through hard work, skill, and creativity (and who thus must build free societies that maximize the human potential of every citizen). We can achieve peace by showing that there is no need to fight for the spoils of war, because it will be made clear that treasure comes not from the ground, but from ourselves.

There is more that will need to be done, but we can do it. With more minds free to participate, we can unleash a global renaissance the like of which has never been seen. We can mobilize our scientific talent and develop controlled fusion, thereby putting to rest all talk of limits. We can use our space program to open the heavens to humanity, offering a grand vision of an infinite future and a soaring salute to the sublime power of the human spirit.

The stars are yet distant. Today the enemy still stands before us: formidable, terrifying, growing his power while he corrupts and cripples ours. Yet we can defeat the blood-drenched religion of fanaticism and hate by faithfully steering a course charted by reason and held to by love of all we hold dear. We can take deliberate action to create the alcohol economy, and, through it, a just global society based on principles of progress and liberty. This is the key to true victory, not merely because such a world will be free of the power of the enemy's oil, but because no one will be able to look upon it and not feel prouder to be human.

POSTSCRIPT

IN DEFENSE OF BIOFUELS

During 2008 world oil prices soared to new heights, with the year-long average expected to run in the range of $120 per barrel. As a result, the United States, which imports 5 billion barrels of oil per year, will face a 2008 petroleum bill that will top $600 billion for foreign oil, plus $400 billion more for domestic oil. This trillion-dollar drain on the economy—equal to 40 percent of the sum that Americans pay in federal taxes—has not only caused record gasoline prices but is also driving the nation into a recession. With less cash available to spend on houses, real estate markets have collapsed, destroying mortgage values and endangering major financial institutions in the process. With less cash available for retail purchases, sales are off on everything from automobiles to cappuccino, forcing businesses to close down or lay off workers, increasing unemployment and cutting government revenues. As a result, the federal deficit is approaching a record $600 billion, and state and local governments will soon be feeling the pinch as well.

And yet, bizarrely, instead of focusing their attention on the staggering cost of oil and its ruinous implications for economic growth and well-being, a host of commentators have begun to decry a different fuel—one that holds the key to ending our dependency on expensive oil purchased from countries with interests inimical to our own.

Biofuels—a class of fuels of which ethanol is the most prominent and immediately promising—can play a central part in weaning the United States off of oil. But starting in early 2008, when the price of oil first crossed the $100-per-barrel threshold, a flood of press reports, articles in scientific journals, and statements from international bureaucrats deluged the media, suggesting that ethanol is starving the world's poor, is a waste of government money, and is bad for the environment. These claims are simply not true; some are based on partial information, some on gross disinformation, but none of them can withstand close scrutiny.

Many of the critics of ethanol may mean well: they are worried about the diversion of food grains from hungry children, or the costs of government biofuel subsidies to taxpayers. Other members of the antiethanol chorus have more self-interested motivations for their criticism of biofuels though, including Saudi oil minister Ali bin Ibrahim Al-Naimi,[1] OPEC president (and Algerian energy minister) Chakib Khelil,[2] and, most vigorously, Venezuelan dictator Hugo Chávez,[3] an ardent proponent of high oil prices who called ethanol production a "crime." Indeed, for those in the know, it hardly came as a surprise when it was revealed that the Glover Park public relations firm, which is registered with the US Department of Justice as an agent of the United Arab Emirates,[4] was heavily involved in orchestrating the antiethanol campaign.[5] Then there are still others, not apparently connected to oil interests, whose opposition to ethanol—and agriculture in general—seems to be driven by a Malthusian vision of a world with fewer people in it. No matter the motivations of these seemingly unlikely bedfellows, their recent objections to ethanol threaten to have the cumulative effect of warping US and international biofuels policy—and just at the moment when exorbitant oil costs should, if anything, be leading legislators to adopt the critical technology needed to expand the role of biofuels in the world's fuel supply.

FUELING RUMORS ABOUT FOOD

Hoping to reduce at least in some small way their need for oil, several countries have adopted energy policies requiring that a percentage of their national fuel supplies consist of biofuels. The European Union, for instance, is aiming to have biofuels make up 10 percent of its vehicle fuel supply by the year 2020. In the United States, legislation in 2005 and 2007 set mandates for ethanol in the nation's fuel mix; the current plan is to ramp up biofuel production until 36 billion gallons are mixed into the nation's fuel supply by 2022.

Unsurprisingly, the result of these mandates has been the rapid expansion of the nation's ethanol industry. The United States, which produced 3 billion gallons of ethanol in 2002, grew its production to 8 billion gallons in 2007, replacing some 5 percent of our gasoline supply.[6] But while this seems like it would be cause for celebration—with enterprising and innovative American farmers helping to reduce our oil usage—some critics have recently alleged that the world's biofuels programs, especially the US corn ethanol effort, are starving poor people around the world by reducing supply and driving up prices. International bureaucrats have been the most vocal critics. A recent World Bank report claimed that "increased biofuel production has contributed to the rise in food prices."[7] The UN's Special Rapporteur on the Right to Food denounced biofuel production as "a crime against humanity."[8] Jeffrey Sachs, a Columbia University economist who is an advisor to UN Secretary-General Ban Ki-moon, has said "we need to cut back significantly on our biofuels programs" because they are "a huge blow to the world food supply."[9] It seems so obvious: With so much corn being turned into fuel, food shortages must inevitably result, and biofuel programs must be the cause.

This conclusion, however, is completely untrue.

Here are the facts. In the last five years, despite the nearly threefold growth of the corn ethanol industry—actually, *because* of it—the amount of corn grown in the United States has vastly increased. The US corn crop grew by 45 percent, the production of distillers grain (a

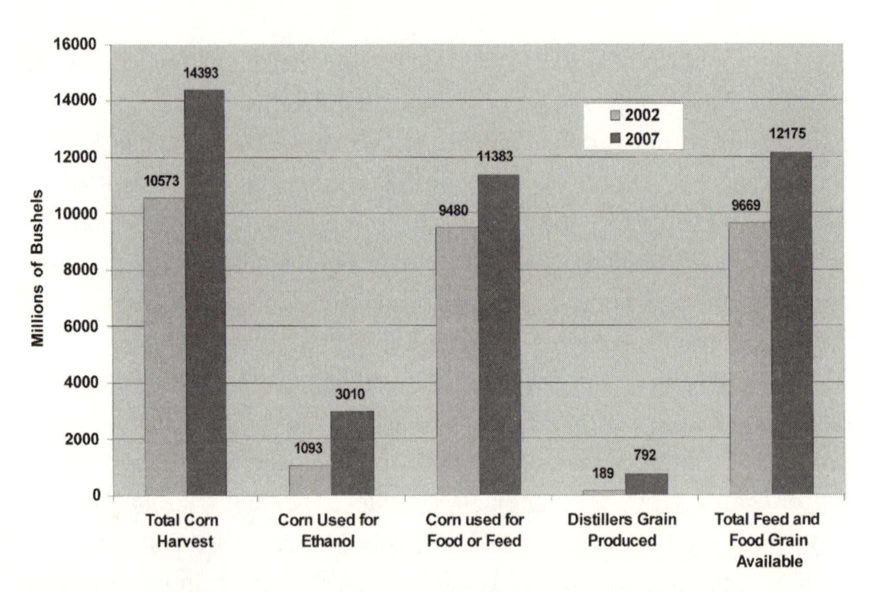

Figure 1. While ethanol production has tripled in the past five years, the amount of grain remaining for food and feed has increased 34 percent. *Source:* Data courtesy of National Corn Growers Association, www.ncga.com.

high-value animal feed made from the protein saved from the corn used for ethanol) quadrupled, and the net US corn product of food for humans and feed for animals increased 34 percent.[10]

Furthermore, contrary to claims that farmers have cut other crops to grow more corn, US soybean plantings this year are expected to be up 18 percent and wheat plantings up 6 percent. US farm exports are up 23 percent over last year.[11] America is clearly doing its share in feeding the world.

At bottom, the entire food-versus-fuel argument boils down to a Malthusian conceit—that there is only so much that can be grown, so if we grow more of one thing we must necessarily grow less of something else. But this is simply false. Agriculture is not a zero-sum game. There are roughly 2,250 million acres of land in the continental United States. About 1,600 million of those acres are arable.[12] Roughly half of that land (800 million acres) is farmland, but only about a third of that (280 million acres) is actually being cultivated.[13] Only about 85 million of those farm acres are presently growing corn, and just a fifth of *that* land—about 17 million acres—is growing corn that becomes ethanol. In short, there is plenty of farmland in the United States that could be used to grow more corn—or more of the other staple crops needed to meet domestic or international demand. Even more important, agricultural technology is constantly advancing. US corn yields per acre have risen 17 percent since 2002, and the state of Iowa alone today produces more corn than the entire nation did in the 1940s. Applied globally, such improved techniques can multiply world agricultural yields many times. In fact, they have risen by a factor of six since 1930—which is why, even though the world's population has tripled since that time, there is a lot more food for everyone today.

So while it is true that there is now much more corn being used for ethanol than ever before, there is also much more total corn than ever before, including much more for food and feed than ever before, and still plenty of land and room for implementation of improved methods to grow even more.

But if biofuels aren't to blame for the rising food prices, what is?

In fact, there are several culprits. One is low farm productivity in some parts of the world. Regional droughts is another. Sometimes there is a confluence of factors: Some critics have foolishly claimed that recent food riots in Haiti could be linked to the US ethanol mandate even though those riots were about rice, which the United States doesn't use to make ethanol, and were largely caused by unwise trade policies and a drought in Australia.

The two primary reasons for higher food prices are, first, higher demand, and, second, higher fuel prices. The increased global demand for food ought to be seen as a very good thing: it represents hundreds of millions of people, especially in China and India, rising out of poverty and affording more calorie-rich diets. Escalating fuel prices, however, are not good news: they drive up the cost of everything we eat. For example, consider the $3 box of cornflakes you might see in

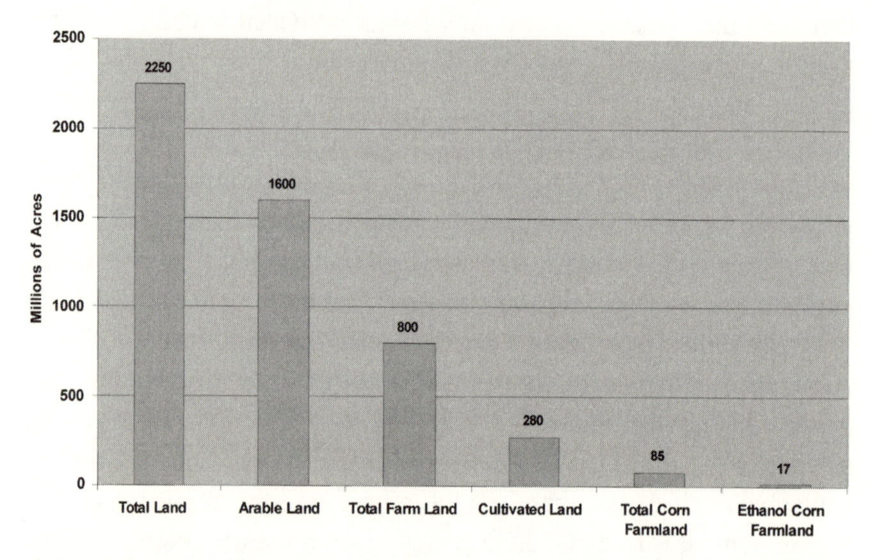

Figure 2. The amount of land being used to grow ethanol is only about 1 percent of the total arable land in the continental United States. *Source:* Data from US Department of Agriculture, www.usda.gov, and the National Corn Growers Association, www.ncga.com.

your grocery store. Farm commodity prices basically have a trivial effect on its price. A bushel of corn contains 56 pounds of grain, so at the current "very high" commodity price of $5 per bushel, a pound of corn costs 9 cents. So the 16 ounces of corn in that cereal box cost a total of 9 cents when bought from the farmer. But when the price of oil goes up, that increases the cost of production, transport, wages, and packaging—all driving up the retail cost of food.

And, in this regard, biofuels have already done more good than harm to the world's poor. According to the *Wall Street Journal* from March 25, 2006, "Global production of biofuels is rising annually by the equivalent of about 300,000 barrels of oil a day. That goes a long way toward meeting the growing demand for oil, which last year rose by about 900,000 barrels a day." The paper cites Merrill Lynch chief commodity analyst Francisco Blanch, who "says that oil and gasoline prices would be about 15 percent higher if biofuel producers weren't increasing their output."[14] So even though the world's biofuels industry is still just aborning, it has already begun to bring down oil prices.

WHY ADAM SMITH WOULD LOVE ETHANOL

That figure from Merrill Lynch contains within it the rebuttal to small-government conservatives and libertarians who believe the United States should give up on ethanol to save money. Those critics generally oppose all government mandates and subsidies and have therefore used the recent antibiofuels push to repeat their long-standing complaints about the federal government's subsidies for biofuels in particular.

However, according to the Merrill Lynch data, the price of oil would be about 15 percent higher were it not for biofuels, which comes to a savings of about $18 per barrel at current oil prices. The United States imports about 5 billion barrels of oil each year. Saving $18 for each barrel adds up to a savings for the country as a whole of $90 billion in foreign oil payments yearly, and a reduction in OPEC global revenues overall of more that $180 billion. This is in addition to cutting another $24 billion from our oil bill by reducing the *amount* of petroleum that we import. Furthermore, the ethanol subsidy saves taxpayers additional money by allowing for the elimination of $8 billion in government-funded crop price supports. Taken together, these savings amount to over $120 billion per year.

This is an extremely good return, considering the pittance that American taxpayers actually shell out for the nation's corn ethanol program: only about $4 billion per year, through a subsidy of 51 cents per gallon.

Nevertheless, many of the libertarian opponents of mandates and subsidies remain troubled by the principle of government interference distorting the markets for food and energy. But it must be remembered: the global markets for food and energy are already badly distorted by trade restrictions in the case of the former, and by the machinations of the OPEC cartel in the case of the latter. Insofar as the nascent biofuels industry will result in eased trade restrictions (so that nations will be able to buy and sell agricultural products for fuel) and in a weakening of OPEC's monopoly power (by bringing into the energy market new fuels that can compete with oil), supporters of free markets should offer three cheers for the rise of biofuels.

It is worth mentioning that Adam Smith, the patron saint of capitalism, was not blindly in favor of markets as the sole determinant of economic activity. Despite his general support for free trade, he wrote in *The Wealth of Nations* that he favored protectionism in cases "when some particular sort of industry is necessary for the defense of the country." After all, "defense is of much more importance than opulence." In fact, he didn't just favor trade restrictions—he even supported subsidies for the sake of national defense: "If any particular manufacture was necessary, indeed, for the defense of the society," Smith wrote, "it might not always be prudent to depend upon our neighbors for the supply; and if such manufacture could not otherwise be supported at home, it might not be unreasonable that all the other branches of industry should be taxed in order to support it."[15] In particular, Smith pointed to the British sailcloth industry—vital to naval propulsion in his day—as eminently deserving of government subsidy. Our need for fuel supplies independent of those imported from unfriendly nations is equally a critical matter of national defense, as Adam Smith would surely note, and thus equally worthy of government support. Defense *is* of much more importance than opulence. A nation that cannot defend itself won't keep its wealth for very long.

LIES, DAMN LIES, AND STATISTICS

For years, the environmental movement supported the US ethanol program on the grounds that, by replacing oil with fuel made from biomass, we can reduce the nation's net emissions of the greenhouse gas carbon dioxide. In 2008, however, the antiethanol campaign struck a forceful blow to undermine this support with a well-publicized study claiming to show the opposite—that the US corn ethanol program actually produces *more* greenhouse gases than would be entailed just by making an equivalent amount of fuel using petroleum, and thus should be condemned by all right-thinking people.

The study, which appeared in the journal *Science*,[16] was authored by a team led by Timothy Searchinger, presently affiliated with Princeton University's Woodrow Wilson School. (Searchinger, it is worth noting, is not a scientist; he is a lawyer who worked, until recently, as a staff attorney for Environmental Defense, the organization best known for the role it played in banning the pesticide DDT in the 1970s—a ban that has resulted in millions of African children dead from malaria.)[17] The Searchinger study offers no new data concerning the US corn ethanol program, conceding—in agreement with numerous previous studies—that the ethanol program's *direct effects* will reduce the nation's greenhouse gas emissions by replacing oil with fuel derived from biomass. However, it then goes on to argue that if *indirect effects* are taken into account, including most notably the potential expansion of third world agriculture in response to the rise of an international market for biofuels, then the overall net effect will be an increase in global greenhouse emissions. Based on a "worldwide agricultural model," the study claims that US agricultural exports will "decline sharply" because more and more American farmland will be used for ethanol—and in order to make up for the lost food supply, Latin American and African peasants will burn down forests to expand farmland. This burning, the study maintains, will put millions of tons of carbon dioxide into the atmosphere, resulting in more emissions than would have come from just burning oil-based fuels.

Not surprisingly, the Searchinger study found instant acclaim from friends of the oil cartel. For example, in its February 13, 2008, issue, the *Wall Street Journal* intoned: "The ink is still moist on Capitol Hill's latest energy bill, and, as if on cue, a scientific avalanche is demolishing its assumptions. To wit, trendy climate-change policies like ethanol and other biofuels are actually worse for the environment than fossil fuels." *Time* magazine also chimed with a sensational article branding biofuels a "scam."[18]

Well, as the saying goes, a lie can circle the globe in the same time it takes truth to put her boots on. While it continues to be cited endlessly in the press, the much-cited study is a Grade A example of junk science.

In the first instance, the real-world data simply do not back up Searchinger's claims. For starters, the study's central assumption—that the rising demand for ethanol has led to a decline in US agricultural exports—is just not true. There has been no reduction in US corn exports, and the US Department of Agriculture projects that corn supplies for food exports, for feed, and for other nonbiofuel uses will continue to grow even as ethanol production expands.[19]

Second, Searchinger's study relies on a flawed assumption about the scope of the US corn ethanol program, one in which the United States will be producing 30 billion gallons of corn ethanol per year by 2015. But in the very same 2007 law that mandated the increased use of biofuels, Congress put a cap on the production of *corn* ethanol—a limit of 15 billion gallons by 2015. This error in the study was pointed out in a devastating online response penned by Michael Wang, a researcher at the Argonne National Laboratory, and Zia Haq, a researcher with the US Department of Energy.[20] Searchinger, they wrote, "examined a corn ethanol production case that is not directly relevant to U.S. corn ethanol production for the next seven years." Wang and Haq's rebuttal is especially powerful since the agricultural model that Searchinger employed was actually first developed by Wang a decade ago.

Third, *contra* Searchinger, there is no evidence that the US corn ethanol program is causing arable land to be cleared elsewhere. To again quote Wang and Haq:

> [Searchinger's assumption about land-use changes] is seriously flawed by predicting deforestation in the Amazon and conversion of grassland into crop land in China, India, and the United States. The fact is, deforestation rates have already declined through legislation in Brazil and elsewhere. In China, contrary to the Searchinger *et al.* assumptions, efforts have been made in the past ten years to convert

marginal crop land into grassland and forest land in order to prevent soil erosion and other environmental problems.

To be clear: Deforestation is certainly happening—and was happening prior to the advent and expansion of the US corn ethanol program. If it *is* accelerating now, that could be due to any number of causes, including, notably, the very high oil prices that the ethanol program serves to combat: The more the global $4 trillion OPEC extortion forces the poor into desperation, the more incentive there will be to cannibalize long-term resources such as forests for lumber or firewood.

And beyond these specific flaws in the study's assumptions, the claim of Searchinger and his colleagues to possess a computer model capable of predicting global human behavior must be taken with a grain of salt. While it might be reasonable to suppose that third world farmers would respond to either high fuel or food prices by clearing more land for agricultural activity, the assumption in the Searchinger study that they would do this by simply burning down their forests—thus creating a "carbon debt" that would take decades or even centuries of biofuel production to "pay back"—is purely speculative. In fact, most of the Amazon deforestation is being driven not by agriculture but by lumbering interests,[21] and should biofuel technology reach the point where either methanol or cellulosic ethanol can be adopted as an economically feasible fuel, then forestry residues would become valuable biofuel resources themselves, and the last thing third world farmers would want to do would be to burn these enormous revenue sources. Instead they would harvest them, and as their energy content would be used to replace petroleum, there would be no significant "carbon debt."

In contrast, it should be noted that a natural ecosystem not subject to human harvesting does little to combat global carbon dioxide levels since its carbon content is in long-term equilibrium with the environment. Thus, if left to itself, a wild forest or grassland will go up in smoke periodically, or otherwise have its biomass decay back into carbon dioxide through microbial action, returning its carbon to the atmosphere with no countervailing petroleum-use reduction benefits.

DO VITAMINS CAUSE GLOBAL WARMING?

Beyond such factual and logical errors, however, the "indirect analysis" methodology used in the Searchinger study is systematically flawed and has nothing in common with the scientific method. Using the same sort of indirect analysis employed by Searchinger—that is, making broad claims of global effects stemming from undemonstrated causal relationships—it is possible to "prove" practically anything. For example, one can also show that increasing mileage standards for vehicles contributes to global warming. Consider: Every gallon of gasoline not used by a motorist saves him $3.50 at today's prices. He can use that money to buy other things. For example, at current prices (about $12 per ton), $3.50 could buy him 560 pounds of coal. Burning that coal would obviously produce far more carbon dioxide emissions than burning the 6 pounds of carbon in one gallon of gas. So higher mileage standards for cars cause global warming, *qed*.

That is an utterly preposterous conclusion, of course, but it exemplifies the Searchinger team's approach. In fact, using indirect analysis, it is possible to show that *any* technology or policy that can be plausibly argued to confer any social benefit whatsoever will cause global warming. For example, both *tax cuts* (because they give consumers greater spending power) and *tax increases* (because they allow for expanded funding of healthcare and public education, which in turn contribute to longer life spans and income growth) can be considered indirect causes of global warming. Perhaps then we should keep taxes the same? Nope—that won't help a bit, since relative to a potential tax cut, level taxes are a tax increase, and relative to a potential tax increase, level taxes are a tax cut. So keeping tax rates the same will cause global warming through both mechanisms—and thus possibly represents the gravest global warming threat of all.

The point isn't simply that the Searchinger study is wrong, but that it represents a method that can be used to produce any conclusion desired. And the desired conclusions, in Searchinger's case, are shaped by what you might call an ethic of "envirostasis"—the belief that the ultimate measure of the merit of any human activity or innovation is its effect on the climate. This way of thinking is profoundly antihuman and can lead rapidly to horrific policy prescriptions.

To see just how pernicious such envirostasis-based ethics can be, let us consider vitamins. Prior to the discovery of vitamins, millions of people—especially poor people with limited diets—were weakened or killed by nutritional deficiencies. But now these people survive, and according to indirect analysis, create a massive global warming effect through their collective carbon footprint. So if vitamins are bad, then it's needless to say that antibiotics are much worse. But even these indirect global warming threats pale before that posed by public sanitation and clean, safe drinking water. Clearly then, according to envirostasis ethics, all such efforts are to be aborted, and medical research, which threatens to bring more such horrors into the world, should be proscribed.

By indirect analysis, however, not only technological innovations but the means of disseminating them also cause global warming. So, for example, the Google search engine, by making technical information much more available to researchers worldwide, must be seen as a major global warming culprit and should be shut down, along with the rest of the Internet. But we must not stop there, because more traditional methods of disseminating information, including books, magazines (even *Science*), newspapers, libraries, the patent registry, and the postal system, would remain as massive global warming agents—as proven by the fact that global warming began before the Internet. So we need to get rid of them too, as well as literacy, just to be sure. Alas, even then, the spoken word would remain, so we might as well get to the heart of the problem—truth. Truth causes global warming.

But it is human reason that discovers truth, creativity that applies it, love that impels its application, and freedom that allows reason, love, and creativity to act. So then, if we are really committed to the protection of climatic stasis, reason must be suppressed, creativity placed in chains, love abandoned, and freedom abolished. To do otherwise is to risk the Fate of the Earth. Vice, on the other hand, is to be celebrated. For it is those who steal their wealth, rather than those who produce it, who have minimal environmental impact; it is those who harm others, who limit their human potential, who constrain their aspirations, and who cut short their hopes, their dreams, and their lives—they are the true heroes of the new creed.

Thus envirostasis thought turns morality on its head. Anything good becomes evil, and all evil becomes good. Searchinger condemns the US corn ethanol program because it allegedly opens up market opportunities for third world peasants. It doesn't—so factually he is wrong—but that is not the problem. *The problem is that his paper opposes the program precisely because of the humanitarian good that it might do.*

So the critical issue at stake ultimately is not ethanol; it's ethics. In their ongoing discourse, the Left and the Right have frequently disagreed on what policies might best advance human well-being, but they nevertheless agree that the advance of human welfare should be the goal of policy. So, for example, the Left might say that minimum wage laws are good for the human condition because they raise the income levels of the poor, while the Right might say they are bad because they cause unemployment; but each side can still break bread with the other, because they both represent rational attempts to reach a common end by alternate means. Republicans might look askance at the universal healthcare plans currently being offered by Hilary Clinton or Barack Obama, but certainly no one can object to the goal of improving healthcare or making it more affordable. Left and Right might disagree on whether public schools or vouchers provide the best educational system, but they both concur that the best educational system should be the goal.

At least that's how it's been up till now. However, if generally accepted, envirostasis ideology provides a radically different way of addressing such problems. Instead of seeking to raise incomes or employment among the poor, we should seek to depress both. Instead of trying to make our healthcare system more effective or affordable, we should do the opposite. Instead of trying to create the best educational system, we should strive for the worst.

And thus we see the ethic of envirostasis revealed for what it really is: rank Malthusian ideology. Conservatives should oppose it for its deeply degrading antihumanism. And liberals, too, should be wary of making common cause with it for the sake of its putative concern about the environment, because all of the proudest accomplishments of both modern and historical liberalism—child labor laws, minimum wage laws, public schools, libraries, urban sanitation, childhood vaccinations, public health services, rural electrification, transportation infrastructure, Social Security, clean air and water laws, civil rights laws, and even slave emancipation, popular enfranchisement, representative government, and independence from colonial rule—all indirectly contribute to carbon emissions, and thus must be ultimately rejected by the cult of envirostasis.

THE REAL ETHANOL ISSUE

Global warming is real. According to well-substantiated measurements, average worldwide temperatures have been increasing for the past several decades at a rate of 0.2 degrees Celsius per decade—a rate that if left unchecked for another century would bring temperatures back where they were a thousand years ago and might raise sea levels approximately one foot. Such a moderate change would hardly be a major threat and could even be beneficial—as it clearly was during the High Middle Ages. However, there is solid reason to believe that this temperature rise is being driven by human carbon dioxide emissions, which are rising as the global economy expands. We will thus need to eventually get carbon dioxide emissions under control. If done properly, replacing petroleum-derived fuels with biofuels could be of great assistance in accomplishing this task, and the evidence suggests that corn ethanol already is making a contribution in that direction.

There is a real flaw in the US corn ethanol program, however, and that is its size: it is much too *small* to effectively address the pressing problem of the looting of our economy by the oil cartel. To put the matter simply: It's not about the weather, it's about the money. The ethanol program is now demonstrably cutting the nation's tribute to the oil cartel by tens of billions of dollars per year. But we need to do much more. The United States paid almost $600 billion for oil imports during FY 2008, an amount taken *out* of the US economy that was more than four times the size of the economic stimulus package Congress authorized in early 2008 to take from the Treasury to put back *into* our economy to stave off recession. Under these circumstances, our nation's modest biofuels program just isn't enough.

We need to do more—and we can. Congress should take the critical step required to break OPEC's vertical monopoly on our economic lifeblood by passing a bill mandating that all new cars sold in the United States be flexible-fueled—that is, able to run on any combination of gasoline, ethanol, or methanol. Such cars already exist and only cost about one hundred dollars more than comparable non-flex-fuel models. By making flex-fuel a requirement for the American auto market, we will make it the international standard as well, and for the first time we will force gasoline to compete at the pump against alcohol fuels all over the world.

Such a flex-fuel vehicle standard would create a global open-source fuel market that would encourage the rise of not only existing sugar and corn ethanol but also other alcohols, including ethanol made from cellulosic material, and methanol, which can be made from any kind of biomass without exception (as well as from coal, natural gas, and even recycled urban trash). By making our cars compatible with such fuels, we will enormously expand and diversify our options, protecting not just Americans but the entire world from domination by the oil cartel.

So long as we do not have fuel choice, the nation will remain at the mercy of the oil cartel, forced to pay any tribute they dictate—whether it's $100 per barrel today or $200 per barrel tomorrow—giving trillions of dollars to Islamists who promote global jihad, fund nuclear weapons development, and take over our corporations and media organizations. But once we open the fuel market, we will put a permanent constraint on the greed and power of our enemies. Indeed, once we have an alternative fuel infrastructure in place, we can defeat them utterly at our pleasure by systematically implementing tax and tariff policies that favor alcohols over oil.

And yes, under those conditions, we will actually create markets for ethanol derived from third world farm products, opening up income opportunities for billions of poor people around the world—just what the envirostasists fear most. We will, in effect, redirect hundreds of billions of dollars from the oil cartel to the world's agricultural sector, creating an enormous engine for global development that will lift whole nations out of poverty.

That will be a very *good* thing to do, and by choosing such a course of action, we will reaffirm human progress as the ethical basis of our society. It will also deliver a powerful rebuke to both the Malthusians and the Islamists, whose common program is not only high oil prices, but the stifling of human initiative and the crushing of human aspirations in order to preserve stasis.

This, and not retreat from the small but promising start the corn ethanol program has made, should be our course. We should not be deterred from it by Malthusian quackery masquerading as science.

GLOSSARY

AROMATICS: Carbon compounds involving rings of carbon atoms. Examples include benzene and toluene. A component of gasoline, aromatics are believed to cause cancer.

ATMOSPHERIC PRESSURE: The pressure an atmosphere exerts. On Earth at sea level, the atmospheric pressure is 14.7 pounds per square inch. This amount of pressure is therefore known as one "atmosphere," or one "bar."

BIOMASS: Combustible material derived from deceased organisms, most typically land or aquatic plants.

BREAKEVEN: The condition wherein a fusion device produces as much energy as is needed to make it run.

BUTANOL: An alcohol that includes four carbon atoms; chemical formula C_4H_9OH. It can be produced from sugar beets.

CATALYST: A substance that facilitates a specific chemical reaction that is not used up in the reaction.

Cetane: A quality rating given to diesel fuel, analogous to the octane rating for gasoline. The higher the cetane rating, the better. Conventional diesel fuel typically has a cetane rating of about 48. DME has a cetane rating of 60.

Corporate Average Fuel Efficiency (CAFE): A set of government standards for increasing the mileage of automobiles.

Cryogenic: Ultracold. Liquid oxygen and hydrogen are both cryogenic fluids, as they require temperatures of –183°C and –253°C, respectively, for storage.

Dhimmi: Islamic term for a Jew, Christian, or Zoroastrian living under Islamic rule. Dhimmis are denied many basic rights, including the right to bear arms or to sue Muslims in court, and so may be brutalized by Muslims at will. They also must pay extra taxes.

Dimethyl ether (DME): A chemical with the formula $(CH_3)_2O$. DME can be produced by reacting methanol with itself. It can be used as diesel fuel or as a raw material for making other chemicals.

E85: A fuel mixture consisting of 85 percent ethanol and 15 percent gasoline.

Electrolysis: The use of electricity to split a chemical compound into its elemental components. Electrolysis of water splits it into hydrogen and oxygen.

Electronic Fuel Injector (EFI): Used to control the feeding of fuel into a modern automobile engine.

Endothermic: A chemical reaction requiring the addition of energy to occur.

Equilibrium constant: A number that characterizes the degree to which a chemical reaction will proceed to completion. A very high equilibrium constant implies near complete reaction.

Ethanol: An alcohol containing two carbon atoms; chemical formula C_2H_5OH. Ethanol is common drinking alcohol in its pure form. It can be made from corn, sugar, grapes, potatoes, and many other crops containing sugar or starch.

Ethylene: A chemical with the formula C_2H_4. Ethylene can be made from petroleum, ethanol, or DME. It is used to make polyethylene, the world's most common plastic.

Exhaust velocity: The speed of the gases emitted from a rocket nozzle. A well-engineered rocket can generally be made to achieve a speed about twice that of its exhaust velocity.

Exothermic: A chemical reaction that releases energy when it occurs.

Fission: A nuclear reaction in which very heavy elements, such as uranium or plutonium, are split to form medium-weight elements, releasing a great deal of energy in the process. Commercial nuclear power plants derive their power from nuclear fission.

Flex-Fuel Vehicle (FFV): A vehicle that can use either alcohol or gasoline, or any mixture of the two, as fuel.

Fuel cell: An electrochemical device that combines hydrogen with oxygen to produce electricity.

Fusion: A nuclear reaction in which light elements, such as isotopes of hydrogen, are fused to form heavier elements, releasing a great deal of energy in the process. The sun and all the stars derive their power from fusion reactions.

General Agreement on Tariffs and Trade (GATT): The international framework set up after World War II to liberalize trade, primarily in manufactured goods. Replaced by the World Trade Organization (WTO) in 1995.

Geothermal energy: Energy produced by using naturally hot underground materials to heat a fluid, which can then be expanded in a turbine generator to produce electricity.

Gigaton (Gt): A billion metric tons.

Greenhouse effect: The action of an atmosphere containing heat-trapping gases such as carbon dioxide to warm the climate of a planet.

Ignition: The condition wherein a fusion machine produces enough energy so that it can run without any energy input. Achieving ignition requires an energy output about four times greater than breakeven (see above).

Integrated Gasification Combined Cycle (IGCC): A new method of making electricity from coal, in which the coal is first reformed with steam into hydrogen and carbon dioxide. The hydrogen is then sent to a gas turbine to generate clean electric power, while the carbon dioxide can be sequestered.

Intergovernmental Panel on Climate Change (IPCC): The UN-sponsored organization concerned with global warming.

Isotope: A variety of a chemical element that has a specific number of neutrons. For example, naturally occurring oxygen always has eight protons in its nucleus, but it can also have either eight neutrons or ten neutrons, with each variety defining a specific isotope, named O-16 and O-18, respectively.

JIZYA: A special ransom tax paid yearly by non-Muslims under Islamic rule in exchange for permission to live.

KELVIN DEGREES: The Kelvin or "absolute" scale is a method of measuring temperature that starts with its zero point set at "absolute zero," the temperature at which a body in fact possesses no heat; 273 K is the same temperature as 0°C, the freezing point of water. Each additional degree Kelvin corresponds to one additional degree centigrade.

KILOCALORIE (KCAL): A unit of energy equal to that delivered by 1 watt in 4,190 seconds.

KILOVOLT (KEV): A unit of plasma temperature frequently used in the fusion program; 1 keV = 11,000,000°C.

KW: Kilowatts.

KWE: Kilowatts of electricity.

KWE-HR: The total amount of energy associated with the use of one kilowatt of electricity for one hour.

M85: A fuel mixture consisting of 85 percent methanol and 15 percent gasoline.

MADRASSA: An Islamic school primarily devoted to teaching the memorization of the Koran. Wahhabi madrassas also teach terrorist ideology.

METHANATION REACTION: A chemical reaction forming methane. For example, hydrogen can be combined with carbon dioxide to produce methane and water.

Methanol: The simplest alcohol, containing only one carbon in its molecule; chemical formula CH_3OH. Also known as "wood alcohol." Methanol can be made from any fuel, biomass, or biomass-derived material, such as wastepaper or other trash.

Mole: The amount of a substance equal in grams to the molecular weight of that substance. For example, water has a molecular weight of 18, so a mole of water would weigh 18 grams.

MWe: Megawatts of electricity.

MWt: Megawatts of heat. One megawatt equals 1,000 kilowatts.

NOx: Air pollutants containing compounds of nitrogen and oxygen such as NO, NO_2, and NO_3.

Organization for Economic Cooperation and Development (OECD): The club of the world's developed nations.

Organization of Petroleum Exporting Countries (OPEC): The international oil cartel consisting of Saudi Arabia, Iran, Iraq, Libya, Algeria, Kuwait, the UAE, Venezuela, Nigeria, Qatar, and Indonesia.

Propanol: An alcohol whose molecule includes three carbon atoms; chemical formula C_3H_7OH. Can be made from propylene.

Propylene: A chemical with formula C_3H_6. Can be made from petroleum or DME. Can be used to make propanol or the important plastic polypropylene.

Pyrolyze: The use of heat to split a compound into its elemental constituents. Methane, CH_4, can be pyrolyzed to produce carbon and hydrogen, for example.

Reverse water gas shift: A chemical reaction in which carbon dioxide is reacted with hydrogen to produce water and carbon monoxide.

Sabatier reaction: A reaction in which hydrogen and carbon dioxide are combined to produce methane and water. The Sabatier reaction is exothermic, with a high equilibrium constant (see above).

Salafism: A school of Sunni Islamic fundamentalism derived from the teaching of the fourteenth-century philosopher Ibn Taymiyyah. Wahhabism is a form of Salafism.

Stable equilibrium: An equilibrium condition that, if displaced by some external force, will return on its own to its original state. A ball on top of a hill is in unstable equilibrium, because if pushed it will roll away, accelerating itself from its original position. A ball in the bottom of a bowl is in stable equilibrium, because if pushed, it will roll back to its starting point.

Steam reformation: The reaction of hot water vapor with a hydrocarbon fuel, such as coal, oil, natural gas, or biomass, to produce a gaseous mixture of carbon monoxide and hydrogen.

Stellarator: An early type of experimental fusion device, now largely supplanted by the tokamak.

Synthesis gas: Also called "syngas." A mixture a carbon monoxide and hydrogen. Syngas is frequently produced by steam reformation of fuels.

Terrawatt (TW): 1,000,000 megawatts. Human civilization today collectively uses about 13 TW.

Tokamak: Named for a Russian acronym meaning "*to*roidal *cha*mber *mag*netic," the tokamak is a kind of experimental fusion device that uses a magnetic field to contain the superhot thermonuclear plasma in a toroidal or doughnut-shaped chamber. Invented in the Soviet Union, the tokamak has outperformed a variety of other devices to become the leading approach to fusion power in all programs worldwide.

TONNE: A metric ton, equal to 2,200 pounds.

TW-YEAR: The total amount of energy associated with the use of one terrawatt for one year. Equal to 3.15×10^{19} joules, or 31.5 exajoules.

UNSTABLE EQUILIBRIUM: An equilibrium that is not stable. See *stable equilibrium*, above.

VAPOR PRESSURE: The pressure exerted by the gas emitted by a substance at a certain temperature. At 100°C, the vapor pressure of water is greater than Earth's atmospheric pressure and so it will boil.

WAHHABISM: An extreme Islamic fundamentalist cult that serves as the state religion of Saudi Arabia.

WATER GAS SHIFT: A chemical reaction in which carbon monoxide is reacted with water to produce carbon dioxide and hydrogen.

NOTES

CHAPTER 1

1. Based upon a compilation of numerous partial reports, an estimate of twenty thousand Saudi-funded madrassas worldwide appears to be *very* conservative. For example, on September 13, 2004, the Saudi royal family made an official announcement of plans to build 4,500 madrassas in south Asia alone. See http://www.indiareacts.com/nati2.asp?recno=2938 (accessed October 10, 2006). In his book *Taliban: Militant Islam, Oil, and Fundamentalism in Central Asia* (New Haven, CT: Yale University Press, 2001), p. 89, respected journalist Ahmed Rashid reports more than 8,000 registered madrassas (and possibly as many as 25,000 unregistered ones) being set up in Pakistan *alone* during the 1980s and 1990s for the purpose of indoctrinating local youth in radical Islamist ideology. According to an online *Asia Times* article (Kaushik Kapisthalam, "Learning from Pakistan's Madrassas," June 24, 2004, http://www.atimes.com/atimes/South_Asia/FF23Df05.html (accessed October 10, 2006), Saudi funding supporting this pre-Taliban programming project exceeds $350 million per year, with a total of more than *2 million* pupils currently enrolled. American correspondent Gerald Posner reports in his book *Secrets of the Kingdom: The Inside Story of the Saudi-US Connection* (New York: Random House, 2005), p. 172, that the Saudis have financed some 210 Islamic centers, 1,500 mosques, 202 colleges, and more

than 2,000 madrassas in *non-Islamic* nations globally. In her online article "Saudi Finance: Sleeping with the Enemy," http://www.monies.cc/publications/saudi_finance.htm (accessed January 10, 2007) the scholarly leftist journalist Loretta Napoleoni reports that 2,715 Wahhabi madrassas were teaching 250,000 students in the Punjab alone in 2001, with the overall Saudi global student indoctrination program funded at a level between $2 and $3 billion per year. In an article in the *Washington Times*, December 10, 2003, available at http://www.benadorassociates.com/article/755 (accessed October 10, 2006), prominent conservative journalist Arnaud de Borchgrave reports 3,000 Saudi-funded madrassas operating in the Philippines, with "tens of thousands . . . spread through Indonesia, Bangladesh, Pakistan, the Middle East, Morocco, sub-Sahara Africa, and North and South America," including some 2,000 in the United States.

2. Gal Luft, "The Future of Oil," Institute for Analysis of Global Security report, http://www.iags.org/futureofoil.html (accessed June 30, 2006).

3. US DOE Energy Information Administration, "International Energy Outlook 2006," http://www.eia.doe.gov/oiaf/ieo/world.html (accessed March 14, 2007).

CHAPTER 2

1. Michael Steinberger, "Lunch with the FT: Bernard Lewis," *Financial Times*, August 9, 2002.

2. Lisa Myers and the NBC Investigative Unit, "Who Are the Foreign Fighters in Iraq? An NBC News Analysis Finds 55 Percent Hail from Saudi Arabia," NBC News, June 20, 2005, http://www.msnbc.msn.com/id/8293410/ (accessed August 4, 2006).

3. Dore Gold, *Hatred's Kingdom* (Washington, DC: Regnery, 2003), pp. 18, 91–92. For additional discussion of Ibn Taymiyyah's thoughts, see Bat Ye'or, *Islam and Dhimmitude: Where Civilizations Collide* (Cranbury, NY: Fairleigh Dickinson University Press, 2002), especially pp. 44, 59.

4. Gold, *Hatred's Kingdom*, pp. 17–26.

5. Ibn Razik, cited in Alexei Vassiliev, *The History of Saudi Arabia* (New York: New York University Press, 2000), p. 106.

6. Gold, *Hatred's Kingdom*, p. 27.

7. Ibid., p. 28.

8. Ibid., p. 34.

9. Said K. Aburish, *The Rise, Corruption, and Coming Fall of the House of Saud* (New York: St. Martin's Press, 1995), p. 24.

10. Daniel Yergin, *The Prize: The Epic Quest for Oil, Money, and Power* (New York: Touchstone Books, 1991), pp. 393–405.

11. Borgna Brunner, ed., *Time Almanac 2007* (Boston, MA: Information Please, 2006), p. 857.

12. Laurent Murawiec, *Princes of Darkness: The Saudi Assault on the West* (Landham, MD: Rowman & Littlefield, 2003), p. 44.

13. Gold, *Hatred's Kingdom*, pp. 54–56, 74–77.

14. Gilles Keppel, *Jihad: The Trail of Political Islam* (Cambridge, MA: Belknap Press, 2003).

15. Gold, *Hatred's Kingdom*, pp. 95–99.

16. Ahmed Rashid, *Taliban: Militant Islam, Oil, and Fundamentalism in Central Asia* (New Haven, CT: Yale University Press, 2001), p. 89. Also see Kaushik Kapisthalam, "Learning from Pakistan's Madrassas," June 24, 2004, http://www.atimes.com/atimes/South_Asia/FF23Df05.html (accessed October 10, 2006), and Loretta Napoleoni, "Saudi Finance: Sleeping with the Enemy," http://www.monies.cc/publications/saudi_finance.htm (accessed January 10, 2007).

17. Ian Black et al., "Militant Islam's Saudi Paymasters," *Guardian*, February 29, 1992.

18. Murawiec, *Princes of Darkness*, p. 47.

19. Wayne Brisard, Jean-Charles Brisard, and Guillaume Dasquie, *Forbidden Truth: U.S.-Taliban Secret Oil Diplomacy and the Failed Hunt for Bin Laden*, trans. Lucy Rounds (New York: Thunder's Mouth Press/Nation Books, 2002), p. 49.

20. Greg Palast and David Pallisser, "FBI and US Spy Agents Say that Bush Spiked Bin Laden Probes before 11 September," *Guardian*, November 7, 2001.

21. Jerry Markon, "U.S. Raids N. Va. Office of Saudi-Based Charity," *Washington Post*, June 2, 2004.

22. Matthew Levitt, "Saudi Financial Counterterrorism Measures: Smokescreen or Substance," *Policywatch*, Washington Institute for Near East Policy, no. 687, December 10, 2002.

23. Gold, *Hatred's Kingdom*, pp. 152–53, 218–19, 243.

24. Murawiec, *Princes of Darkness*, p. 97.

25. Alexei Alexiev, "The Missing Link in the War on Terror: Confronting Saudi Subversion," *National Review*, October 28, 2002.

26. US Department of State, "Report of the Accountability Review Boards of the Embassy Bombings in Nairobi and Dar es Salaam on August 7, 1998," January 1999, http://www.state.gov/www/regions/africa/accountability_report.html (accessed September 6, 2006).

27. Jeff Gerth and Judith Miller, "Saudis Called Slow to Help Stem Terror Finances," *New York Times*, December 1, 2002.

28. Gold, *Hatred's Kingdom*, pp. 151–52.

29. David Kaplan, Monica Ekman, and Amir Latif, "The Saudi Connection: How Billions in Oil Money Spawned a Global Terror Network," *U.S. News & World Report*, December 15, 2003, http://foi.missouri.edu/terrorbkgd/saudiconnect.html (accessed September 15, 2006).

30. An official list of such Saudi-funded Islamic centers established worldwide has been published by the Saudi Arabian government at http://www.kingfahdbinabdulaziz.com/main/m400.htm (accessed October 10, 2006).

CHAPTER 3

1. Robert Baer, *Sleeping with the Devil: How Washington Sold Our Souls for Saudi Crude* (New York: Three Rivers Press, 2003), p. 43.

2. Contingency plans for such a seizure had been drawn up by the Pentagon, which estimated that two brigades would be sufficient: one for Saudi Arabia and the other for Kuwait. According to a 1973 military planning document recently made public by the *Washington Post*, in Saudi Arabia the US forces would have had to face only one "lightly armed national guard battalion at Dharan" and a US-made Hawk surface-to-air-missile battery. See Glenn Frankel, "US Mulled Seizing Oil Fields in '73," *Washington Post*, January 1, 2004, p. 1.

3. US Department of Justice, "Report of the Attorney General to the Congress of the United States on the Administration of the Foreign Agents Registration Act of 1938, as Amended, for the Six Months Ending June 30, 2005," 2005, http://www.usdoj.gov/criminal/fara/reports/1sthalf2005FARAReporttoCongress.pdf (accessed July 15, 2006).

4. See Loeffler's biography at his firm's Web site: http://www.loefflerllp.com/LTPR/ourpeople_1.asp?id=57 (accessed February 12, 2007).

5. See "Kissinger Quits 9/11 Panel," CBS News, December 13, 2002, http://www.cbsnews.com/stories/2002/12/14/terror/main533049.shtml (accessed June 12, 2006). See also Dan Eggen, "Kissinger Quits Post as Head of 9/11 Panel," *Washington Post*, December 14, 2002, p. 1, http://www.washingtonpost.com/ac2/wp-dyn?pagename=article&node=&contentId=A52655–2002Dec13¬Found=true (accessed June 12, 2006); and David Corn, "Probing 9/11," *Nation,* July 7, 2003, http://www.thenation.com/doc/20030707/corn (accessed June 12, 2006).

6. Aram Roston, "A Royal Scandal," *Nation*, December 3, 2001, http://www.thenation.com/doc/20011203/roston (accessed July 10, 2006). Also see Baer, *Sleeping with the Devil*, pp. 151–56.

7. Baer, *Sleeping with the Devil*, pp. 46–50.

8. Ibid., p. 51. See also David Ignatius, "Bush's Fancy Financial Footwork," *Washington Post*, August 6, 2002, p. 15, http://www.washingtonpost.com/ac2/wp-dyn/A48301–2002Aug6 (accessed July 11, 2006).

9. Michael Isikoff and Mark Hosenball, "A Legal Counterattack: Saudis Hire Some of the Toniest U.S. Law Firms to Defend Them against the Landmark $1 Trillion Lawsuit on Behalf of the Victims of 9-11," *Newsweek*, April 16, 2003, http://www.msnbc.msn.com/id/3067906/. Most recently, Baker made headlines as leader of the Baker-Hamilton Iraq Study Group (ISG). The ISG made seventy-nine recommendations, many of which, such as the redeployment of US forces in Iraq from a frontline combat to a predominantly training role, were plausible, if controversial. Others, such as its call for the United States to pressure Israel to withdraw from the Golan Heights as a way to reduce terrorism in Baghdad, were absurd, and provoked much discussion about various rude anti-Israel and anti-Jewish comments that Baker had reportedly made in the past. The greatest defect in the ISG report, however, was its failure to mention a need for the United States to pressure Saudi Arabia to stop encouraging its subjects to engage in acts of terrorism against US forces and peaceful civilians in Iraq. This was a glaring omission, since it has been documented (see chap. 2, n.2) that the majority of foreign terrorists committing the worst acts of violent mayhem in Iraq are Saudis. Akin Gump Senior Counsel Vernon Jordan was also a member of the ISG. For a roster of many others with Saudi connections who participated in the ISG, see Ed Lasky, "Baker's ISG: Shilling for the Saudis," *American*

Thinker, December 19, 2006, http://www.americanthinker.com/2006/12/personnel_is_policy_the_case_o.html (accessed April 1, 2007).

10. Baer, *Sleeping with the Devil*, p. 54.

11. Ibid., p. 60.

12. See Sultan's biography at http://en.wikipedia.org/wiki/Sultan%2C_Crown_Prince_of_Saudi_Arabia (accessed March 20, 2007).

13. Baer, *Sleeping with the Devil*, p. 57.

14. Jonathan Wells, Jack Meyers, and Maggie Mulvihill, "Bush Advisors Cashed in on Saudi Gravy Train," *Boston Herald*, December 11, 2001, http://www.commondreams.org/headlines01/1211–05.htm (accessed March 20, 2007).

15. Baer, *Sleeping with the Devil*, p. 58.

16. See "Akin Gump Strauss Hauer & Feld," wikipedia.org, http://en.wikipedia.org/wiki/Akin_Gump_Strauss_Hauer_&_Feld (accessed March 31, 2007).

17. See "U.S. Targets Assets of Suspected Hamas Financiers: Officials Warn Financial Strike May Be Just the Beginning," CNN, December 4, 2001, http://archives.cnn.com/2001/US/12/04/inv.bush.terror/index.html (accessed March 30, 2007). Also, Maggie Mulvihill, Jonathan Wells, and Jack Meyers, "White House Connections: Saudi Agents Close Bush Friends," *Boston Herald*, December 11, 2001, http://socrates.berkeley.edu/~pdscott/q4c.html (accessed March 31, 2007).

18. Baer, *Sleeping with the Devil*, p. 57.

19. Laurent Murawiec, *Princes of Darkness: The Saudi Assault on the West* (Lanham, MD: Rowman & Littlefield, 2005) pp. 51–53. Also see Gerald Posner, *Secrets of the Kingdom: The Inside Story of the Saudi-U.S. Connection* (New York: Random House, 2005), p. 88.

20. Joseph Farah, "Prince Al Waleed Bin Talal and the Media," *World Net Daily*, November 7, 2001, http://www.freelebanon.org/articles/a198.htm (accessed March 31, 2007).

21. Frank Gaffney, "Fox's Saudi Prince," *Front Page*, September 30, 2005, http://www.frontpagemag.com/Articles/ReadArticle.asp?ID=19652 (accessed March 31, 2007). See also Wes Vernon, "Radical Arabs Influence over US Media," Accuracy in Media, December 6, 2005, http://www.aim.org/aim_report/4220_0_4_0_C/ (accessed March 31, 2007).

22. Eliana Johnson, "Ties to Terrorism? No Thanks, Alwaleed." *Yale Daily News*, January 17, 2006, http://www.yaledailynews.com/articles/view/16079?badlink=1 (accessed March 31, 2007).

23. "Saudi Billionaire Boasts of Manipulating Fox News Coverage," Accuracy in Media, December 7, 2005, http://www.aim.org/press_release/4222_0_19_0_C (accessed April 2, 2007).

24. Baer, *Sleeping with the Devil*, pp. 23–25, 27–28, 76, 163–64. I have also spoken with a very famous American who has been to some of these parties. Apparently, for those of a certain sort, they are worth flying halfway around the world to attend.

25. Ibid., p. 66.

26. Ibid., p. 64.

27. Steven Emerson, *The American House of Saud: The Secret Petrodollar Connection* (New York: Franklin Watts, 1985), p. 373. Also see Murawiec, *Princes of Darkness*, p. 134; and Baer, *Sleeping with the Devil*, pp. 64–66.

28. David Paul Kuhn, "The Tangled Web of US-Saudi Ties," CBS News, April 20, 2004, http://www.cbsnews.com/stories/2004/04/20/politics/main612852.shtml (accessed March 31, 2004).

29. Michael Isikoff and Evan Thomas, "The Saudi Money Trail," *Newsweek*, December 2, 2002. Synopsis online at http://watch.windsofchange.net/themes_28b.htm#trail (accessed March 31, 2007).

30. Daniel Pipes, "Government for Sale [to the Saudis]," *New York Post*, December 3, 2002, http://www.danielpipes.org/article/980 (accessed March 31, 2007).

31. Rachel Ehrenfeld, "A 'Political Party' Unveiled," *Washington Times*, August 11, 2006, http://www.washtimes.com/op-ed/20060810–084245–5230r.htm (accessed March 31, 2007). Also see "Proxy Groups," Global Security, http://www.globalsecurity.org/intell/world/lebanon/proxy-groups.htm (accessed March 31, 2007).

32. Toby Westerman, "Terrorists Active in US Backyard: Latin America a Hotbed for Both al-Qaida, Hezbollah," *World Net Daily*, May 7, 2002, http://www.worldnetdaily.com/news/article.asp?ARTICLE_ID=27521 (accessed March 31, 2007).

33. Yuliya Tymoshenko, "Moscow's Mideast Myopia," *Haaretz.com*, January 17, 2007, http://www.haaretz.com/hasen/spages/813794.html (accessed March 31, 2007). See also Pavel Felgenhauer, "Who Will Be Russia's Best Friend," *Perspective* 8, no. 1 (September–October 2002), http://www.bu.edu/iscip/vol13/felgenhauer.html (accessed March 31, 2007).

34. For example, see "Thousands Flee Uzbekistan as the Uprising

Reaches Other Towns," *Pravda-RU*, May 16, 2005. Available online in English at http://english.pravda.ru/hotspots/terror/8248–0/ (accessed March 31, 2007). For an anecdotal account of a tour through such violence, see Robert Baer, *See No Evil: The True Story of a Ground Soldier in the CIA's War on Terrorism* (New York: Three Rivers Press, 2002) pp. 141–67.

35. David Filipov, "Drug Addiction and HIV Infection Soars in Russia," *Boston Globe*, February 27, 1999, http://www.psychosocial.com/addiction/drugruss.html (accessed April 1, 2007).

36. "Gulf War," *Encyclopaedia of the Orient*, http://lexicorient.com/e.o/gulfwar.htm (accessed March 31, 2007).

37. James A. Phillips, "What George Bush Now Must Do in Iraq," Executive Memorandum #300, Heritage Foundation, April 4, 1991, http://www.heritage.org/Research/MiddleEast/EM300.cfm (accessed July 10, 2006).

38. Rachel Ehrenfeld, *Funding Evil: How Terrorism Is Financed—and How to Stop It* (Chicago: Bonus Books, 2003).

39. Rachel Ehrenfeld, "Down and Out in Palestine," *Washington Times*, March 15, 2001. Also see appendix B in National Criminal Intelligence Service (NCIS), "An Outline Assessment of the Threat and Impact of Organised/Enterprise Crime upon United Kingdom Interests," UK government briefing paper, 1993.

40. Ehrenfeld, *Funding Evil*, p. 72.

41. Baer, *Sleeping with the Devil*, p. 66.

42. Ibid., p. 158.

43. Robert D. McFadden, "Looking for an Attorney General: The Dispute; the White House and Judges Allies Clash over Hiring," *New York Times*, February 7, 1993, http://select.nytimes.com/search/restricted/article?res=F00616F8355E0C758CDDAB0894DB494D81 (accessed March 31, 2007). Wood's detractors compared her to Zoe Baird, a previously discarded candidate who had broken the law by hiring two illegal immigrants and not paying their taxes. However, the cases were not comparable, because Wood had hired her nanny with an expired visa in 1986, when the practice was entirely legal, and both Wood and the employee had paid taxes and filed all required paperwork with the INS, the IRS, and other state and federal authorities. Furthermore, shortly after being hired the worker in question had her documentation remedied and become a legal resident and, subsequently, a US citizen. The "scandal" was thus a purely synthetic political smear job.

44. Warwick Sabin, "Coffee with Clinton's Consligiere: Bruce Lindsey

Talks about Life with the 42nd President," *Arkansas Times*, November 11, 2004, http://www.arktimes.com/Articles/ArticleViewer.aspx?ArticleID=821ad8eb-3c09–4915-a846-acf79cc9dd08 (accessed April 1, 2007). There is no proof that Lindsey's appointment at Akin Gump was a payoff for blocking Wood's nomination. It is possible that the sequence of events was coincidental. However, whenever a former public official receives a reward from an entity representing any private interest—let alone a foreign government or criminal organization—for whom he or she had done favors while holding government authority, it should raise questions. The fact that leading Washington lobbying organizations that provide well-paid "revolving door" sinecures for public officials when out of office have been allowed to accept large sums of money from the Saudi government and indicted terrorist-linked clients thus poses a serious threat to the integrity of the United States government. People like Khalid bin Mahfouz should not have a voice in choosing the US attorney general.

45. Gerald Posner, *Why America Slept: The Failure to Prevent 9/11* (New York: Ballantine Books, 2003). See also Heather MacDonald, "Why the FBI Didn't Stop 9/11," *City Journal*, August 2002, http://www.city-journal.org/ html/12_4_why_the_fbi.html (accessed April 1, 2007); and "Panel: FBI Lost Chance to Stop 9/11: Sept. 11 Commission Says the Agency Failed to Detect Key Al Qaida Cell," Associated Press, April 14, 2004, http://media.www.michigandaily.com/media/storage/paper851/news/2004/04/14/News/Panel.Fbi.Lost.Chance.To.Stop.911–1423857.shtml (accessed April 1, 2007).

46. Posner, *Why America Slept*, pp. 60, 70–71, 73; Baer, *Sleeping with the Devil*, p. 35; and J. Michael Waller, "Policy Disaster," *Insight on the News*, November 13, 2000, http://www.findarticles.com/p/articles/mi_m1571/is_42_16/ai_72328780/pg_1 (accessed April 1, 2007).

47. Saxby Chambliss, "Statement of Representative Saxby Chambliss, Chairman, House Intelligence Subcommittee on Terrorism and Homeland Security before the House Armed Services Committee Special Oversight Panel on Terrorism," September 5, 2003, http://www.fas.org/irp/congress/2002_hr/090502chambliss.html (accessed April 1, 2007). See also House Report 108–561 for FY 2005, Section 304, "The 1995 'Deutch Guidelines' regarding the recruitment of foreign assets impeded human intelligence collection efforts and contributed to the creation of a risk averse environment. Despite repeated efforts by the intelligence oversight committees of Con-

gress to convince the Director of Central Intelligence to drop the guidelines, these guidelines stood until formally repealed in 2001 by an Act of Congress," http://thomas.loc.gov/cgi-bin/cpquery/?&sid=cp108cT0f3&refer=&r_n=hr561.108&db_id=108&item=&sel=TOC_33747& (accessed April 1, 2007).

48. See the entry for McLarty at the Carlyle Group Web site, http://www.thecarlylegroup.com/eng/team/l5-team2741.html (accessed April 1, 2007).

49. Baer, *Sleeping with the Devil*, p. 167.

50. Murawiec, *Princes of Darkness*, pp. xi–xxi.

51. Craig Unger, *House of Bush, House of Saud: The Secret Relationship between the World's Two Most Powerful Dynasties* (New York: Scribner, 2004).

52. "Bin Laden Family Evacuated," CBS News, September 30, 2001, http://www.cbsnews.com/stories/2001/09/30/archive/main313048.shtml (accessed April 1, 2007).

53. The fact that the genocide in Darfur is being perpetrated by the Sudanese government, and not by autonomous forces, is well known. See Human Rights Watch, "Sudan: Government Commits Ethnic Cleansing in Darfur," *Human Rights News*, May 7, 2004, http://hrw.org/english/docs/2004/05/07/darfur8549.htm (accessed April 1, 2007). Also see Glenn Kessler and Colum Lynch, "U.S. Calls Killings in Sudan Genocide: Khartoum and Arab Militias Are Responsible, Powell Says," *Washington Post*, September 10, 2004, p. 1, http://www.washingtonpost.com/wp-dyn/articles/A8364–2004Sep9.html (accessed April 1, 2007).

54. Peter S. Goodman, "China Invests Heavily in Sudan's Oil Industry: Beijing Supplies Arms Used on Villagers," *Washington Post*, December 23, 2004, p. 1, http://www.washingtonpost.com/wp-dyn/articles/A21143–2004Dec22.html (accessed April 1, 2007).

CHAPTER 4

1. The actual quote is "To wage war, three things are necessary: money, money, and yet more money." It was reportedly said in 1499 by Gian-Jacopo Trivulzio, marshal of France, when asked by King Louis XII what he needed for a campaign in Italy.

2. "Saudi Oil Policy: Stability with Strength," speech by Minister of Petroleum and Mineral Resources Ali Al-Naimi, Houston, TX, October 20, 1999. Available at the Saudi Arabian embassy Web site at http://www.saudiembassy.net/1999News/Statements/SpeechDetail.asp?cIndex=327 (accessed April 3, 2007).

3. Raymond Learsy, *Over a Barrel: Breaking the Middle East Oil Cartel* (Nashville, TN: Nelson Current, 2005).

4. Data from the US DOE Energy Information Agency (EIA). Available online at http://tonto.eia.doe.gov/dnav/pet/pet_pri_spt_s1_m.htm (accessed April 3, 2007).

5. Learsy, *Over a Barrel*, p. 146.

6. Different authorities use different criterion as to how to count oil production, which is why reports of total current world oil production vary between 72 and 85 million barrels of oil per day. WTRG Economics, for example, excludes from its count volatile distillates such as ethane and propane, which outgas from the petroleum after brought to the surface, and derives its output statistics from measurements made after certain other processing steps that reduce the petroleum product volume have also occurred. Its production figures thus correspond to the low end of the spectrum.

7. Learsy, *Over a Barrel*, p. 233.

8. Borgna Brunner, ed., *Time Almanac 2006* (Boston MA: Information Please, 2005) p. 804.

9. Ibid., p. 711.

10. Learsy, *Over a Barrel*, pp. 4–5.

11. Amy Myers Jaffe and Robert A. Manning, "The Shocks of a World of Cheap Oil," *Foreign Affairs*, January–February 2000.

12. Donella H. Meadows, *The Limits to Growth: A Report for the Club of Rome's Project on the Predicament of Mankind* (New York: Macmillan, 1974). In the years since its publication, this classic Malthusian prophesy of doom has been shown to be wrong on every point.

13. Julian Simon, *The Ultimate Resource 2* (Princeton, NJ: Princeton University Press, 1996). This is an update of an (incorrectly) much-derided 1970s work by Simon in which he argued (correctly) that all the predictions of the Club of Rome would be proven wrong.

14. Peter Tertzakian, *A Thousand Barrels a Second: The Coming Oil Break Point and the Challenges Facing an Energy Dependent World* (New York: McGraw-Hill, 2006).

15. For a sampling of Lovins's ideas, both practical and impractical, see Amory Lovins et al., *Winning the Oil Endgame* (Snowmass, CO: Rocky Mountain Institute, 2004).

16. Robert Bamberger, "Automobile and Light Truck Fuel Economy: The CAFE Standards," Congressional Research Service, January 13, 2002, http://www.csa.com/discoveryguides/ern/03aug/IB90122.php (accessed April 3, 2007).

17. Tertzakian, *A Thousand Barrels a Second*, p. 111.

18. China's oil imports hit 4 million barrels per day in 2006. At its current rate of growth of 10 percent per year, it will reach 10 million barrels per day in 2016, effectively putting another customer the size of the United States into the world oil market. India will not be far behind. These developments promise to drive the price of oil through the roof.

19. Tertzakian, *A Thousand Barrels a Second*, p.117–18.

20. Ibid., pp. 88–90.

21. Statistics from EIA Web site, http://www.eia.doe.gov/emeu/aer/elect.html (accessed April 3, 2007). Note that I have translated their units of billion kilowatt hours per year to more straightforward units of million kilowatts.

CHAPTER 5

1. Energy Information Agency (EIA) oil statistics, available online at http://www.eia.doe.gov/emeu/cabs/topworldtables3_4.html and http://www.eia.doe.gov/emeu/cabs/topworldtables1_2.html (accessed April 3, 2007). The EIA is a division within the US Department of Energy.

2. Borgna Brunner, ed., *Time Almanac 2006* (Boston, MA: Information Please, 2005), p. 585. Additional figures are available from the EIA Web site http://www.eia.doe.gov/emeu/international/oilreserves.html (accessed April 3, 2007).

3. "The Future of Oil," Institute for the Analysis of Global Security, 2003, http://www.iags.org/futureofoil.html (accessed April 3, 2007). The IAGS report in turn drew its data from the BP Statistical Review of World Energy.

4. EIA coal production and reserve data, available online at http://www.eia.doe.gov/fuelcoal.html (accessed April 3, 2007).

5. EIA natural gas production data, available online at http://www.eia.doe.gov/emeu/international/gasproduction.html (accessed April 3, 2007).

6. D. O. Hall and K. K. Rao, *Photosynthesis* (Cambridge: Cambridge University Press, 1999); World Energy Council, "WEC Survey of Energy Resources 2001—Biomass," 2001, http://www.worldenergy.org/wec-geis/publications/reports/ser/biomass/biomass.asp (accessed April 3, 2007).

7. Data taken from UN Food and Agricultural Organization (FAO) Web site, http://faostat.fao.org/site/339/default.aspx (accessed April 4, 2007).

8. Wang Mengjie and Ding Suzhen, "A Potential Renewable Energy Resource Development and Utilization of Biomass Energy," paper no. 9408, Chinese Academy of Agricultural Engineering Research and Planning, Beijing, China, 1994, http://www.fao.org/docrep/T4470E/t4470e0n.htm (accessed April 4, 2007). Note that I have converted their units of exajoules of biomass to metric tonnes. An exajoule (10^{18} joules) is the energy equivalent of about 68 million metric tonnes of dry biomass. The data in table 5.6 should be considered conservative, as it dates from 1987. As global crop yields have increased by about 50 percent since that time, agricultural waste probably has as well. A much more recent study of the global potential of crop residues, but only as they relate to ethanol crops, is Seungdo Kim and Bruce Dale, "Global Potential Bioethanol Production from Wasted Crops and Crop Residues," *Biomass and Bioenergy* 26 (2004): 361–75, http://www.icpf.cas.cz/trogl/download/bioethanol/uroda.pdf (accessed April 3, 2007). The authors show that expendable crop residues alone can replace one-third of the global gasoline supply with ethanol.

9. George Olah, Alain Goeppert, and G. K. Surya Prakash, *The Methanol Economy* (Weinheim, Germany: Wiley-VCH, 2006) p. 114.

CHAPTER 6

1. Roberta Nichols, "The Methanol Story: A Sustainable Fuel for the Future," *Journal of Scientific and Industrial Research* 62 (January–February 2003): 97–103.

2. Ibid., p. 99.

3. Richard Wineland, Roberta Nichols, and Eric Clinton, "Control system for engine operation using two fuels of different volumetric energy content," US Patent 4,706,629, issued November 17, 1987.

4. Richard Wineland, Roberta Nichols, and Eric Clinton, "Spark timing control of multiple fuel engine," US Patent 4,703,732, issued November 3, 1987.

5. Richard Wineland, Roberta Nichols, and Eric Clinton, "Control system for engine operation using two fuels of different volatility," US Patent 4,706,630, issued November 17, 1987.

6. Tom MacDonald, "California's Twenty Years of Alcohol-Fueled Vehicle Demonstrations Reviewed," paper presented at the Thirteenth International Symposium on Alcohol Fuels, Stockholm, Sweden, 2000.

7. "Methanol Health Effects Fact Sheet," Methanol Institute, available at www.methanol.org (accessed April, 27, 2006).

8. George Olah, Alain Goeppert, and G. K. Surya Prakash, *The Methanol Economy* (Weinheim, Germany: Wiley-VCH, 2006).

9. Nichols, "The Methanol Story," pp. 97–103.

10. The reaction is $2CH_3OH \rightarrow (CH_3)_2O + H_2O$.

11. Olah, Goeppert, and Prakash, *The Methanol Economy*, p. 204. Citations from *The Methanol Economy*: W. H. Cheng, H. H. Kung, eds., *Methanol Production and Use* (New York: Marcel Dekker, 1994); "Methanol Fuels and Fire Safety," Fact Sheet OMS-8, EPA 400-F-92–010, US Environmental Protection Agency (EPA), Office of Mobile Sources, 1994.

CHAPTER 7

1. George Olah, Alain Goeppert, and G. K. Surya Prakash, *The Methanol Economy* (Weinheim, Germany: Wiley-VCH, 2006), p. 153.

2. F. David Doty, "Fuel for Tomorrow's Vehicles," Doty Scientific Report, August 16, 2004, www.accstrategy.org/workingpapers/dotyfuturefuels .pdf (accessed April 6, 2007). See also data available at the Fuel Cell Store, 2004, http://www.fuelcellstore.com/cgu-bin/fuelweb/view=subcat/cat=23/ subcat=27 (accessed July 23, 2006).

3. Ballard Power financial data available at http://finance.yahoo.com/ q?s=bldp&d=t,2004 (accessed July 23, 2006). See also Joseph J. Romm, *The Hype about Hydrogen: Fact and Fiction in the Race to Save the Climate* (Washington, DC: Island Press, 2004), pp. 115–24.

4. Doty, "Fuel for Tomorrow's Vehicles."

5. Committee on Alternatives and Strategies for Future Hydrogen Production and Use, National Academy of Engineering, *The Hydrogen Economy: Opportunities, Costs, Barriers, and R&D Needs* (Washington, DC: National Academies Press, 2004.

6. James J. Eberhardt, "Fuels of the Future for Cars and Trucks," paper presented at DEER 2002, www.osti.gov/fcvt/deer2002/deer2002wkshp.html (accessed September 8, 2006).

7. James K. Glassman and Kevin A. Hassett, *Dow 36,000: The New Strategy for Profiting from the Coming Rise in the Stock Market* (New York: Three Rivers Press, 1999).

8. Kevin A. Hassett, "Ethanol's a Big Scam, and Bush Has Fallen for It," *Bloomberg News Service*, February 13, 2006.

9. D. Pimentel, "Ethanol Fuels: Energy Balance, Economics, and Environmental Impacts Are Negative," *Natural Resources and Research* 12, no. 2 (2003): 127–34. Also see D. Pimentel and T. W. Patzek, "Ethanol Production Using Corn, Switchgrass, and Wood; Biodiesel Production Using Soybean and Sunflower," *Natural Resources Research* 14, no. 1 (2005): 65–76.

10. S. Kim and B. Dale, "Environmental Aspects of Ethanol Derived from No-Till Corn Grain: Nonrenewable Energy Consumption and Greenhouse Gas Emissions," *Biomass and Bioenergy* 28 (2005): 475–89. See also S. Kim and B. Dale, "Allocation Procedure in Ethanol Production from Corn Grain," LCA Case Studies, *International Journal of Life Cycle Analysis* 7, no. 4 (2003).

11. A. Farrell, R. Plevin, B. Turner, A. Jones, M. O'Hare, and D. Kammen, "Ethanol Can Contribute to Energy and Environmental Goals," *Science* 311 (January 27, 2006): 506–508.

12. Bjorn Lomberg, *The Skeptical Environmentalist* (Cambridge: Cambridge University Press, 2002). See pp. 21–27, 104–105, 216, 246, and 251, in the course of which Lomberg refutes a string of papers by Pimentel.

13. David Pimentel et al., "Ecology of Increasing Disease: Population Growth and Environmental Degradation," *Bioscience* 48, no. 10 (1998): 817–26.

14. "Ecologist Says Unchecked Population Growth Could Bring Misery," American News Service, 1998, http://www.utne.com/web_special/web_specials_archives/articles/799–1.html (accessed September 14, 2006).

15. Here's a sample: "Pimentel says that in order for every person on earth to have adequate resources of food, shelter and clothing, the ideal pop-

ulation on the earth should be about 2 billion—approximately the number of people living on the planet in the 1950s. These fortunate 2 billion will be free from poverty and starvation, living in an environment capable of sustaining human life with dignity, the report suggests. But even at a reduced world population—achieved, ideally, by democratically determined population control practices and sound resource-management policies—life for the average person cannot be as luxurious as it is for many Americans today." David Pimentel et al., "Will the Limits of the Earth's Resources Constrain Human Numbers?" *Environmental Development, and Sustainability* 1 (1999): 19–39. Other Pimentel writings on the need to slash the world's population include David Pimentel and Marcia Pimentel, "Land, Energy and Water: The Constraints Governing Ideal U.S. Population Size," Negative Population Growth Forum, 1990, http://dieoff.org/page136.htm (accessed April 9, 2007); David Pimentel, Xuewen Huang, Ana Cordova, and Marcia Pimentel, "Impact of Population Growth on Food Supplies and Environment," paper presented at the American Academy for the Advancement of Science Annual Meeting, February 9, 1996; and David Pimentel, R. Harman, M. Pacenza, J. Pecarsky, and M. Pimentel, "Natural Resources and an Optimum Human Population," *Population and Environment* 15, no. 5 (1994): 347–69.

16. Stuart Anderson, "The World According to Dick Lamm," Cato Institute, July 28, 1996, http://www.cato.org/dailys/7–28–96.html (accessed April 9, 2007); Thomas Sowell, "The Duty to Die," *Jewish World Review*, April 26, 2001, http://www.jewishworldreview.com/cols/sowell042601.asp (accessed April 9, 2007).

17. For Sierra Club Executive Director Carl Pope's warning about the racist hate groups, see "A Message from Carl Pope," http://www.alleghenysc.org/article.html?itemid=200403061101440.199409 (accessed September 19, 2006). For a sample of the response of some of Pimentel's supporters, see "Save the Sierra Club from Homo Jew Takeover," available online at http://www.overthrow.com/lsn/news.asp?articleID=6541 (accessed September 19, 2006). Overthrow.com calls itself "the official website of the Libertarian Socialist Party." Their Web site, which features a photo of Heinrich Himmler on its front page, proudly announces that "Heinrich Himmler is the patron saint of the Libertarian Socialist Party."

18. Fred Pearce, "The Greening of Hate," interview with Betsy Hartmann, *New Scientist*, February 20, 2003, http://www.hartford-hwp.com/archives/25b/027.html. See also the interesting discussion of this article at http://peakoildebunked.blogspot.com/2005/09/106-greening-of-hate.html

and further discussion of Pimentel's questionable connections at http://peakoildebunked.blogspot.com/2005/10/123-more-racist-connections.html (all accessed September 19, 2006).

19. Byron Wells, "Migrant Foe Tied to Racism," *East Valley Tribune*, August 1, 2004, http://www.eastvalleytribune.com/?sty=26431 (accessed April 9, 2007). See also Southern Poverty Law Center, "White Supremacy: Ignoring Its Own Ties, Anti-immigrant Group Denounces White 'Separatist,'" *Intelligence Report*, Fall 2004, http://www.splcenter.org/intel/intelreport/article.jsp?aid=498 (accessed April 9, 2007); "Extremist Leads New Arkansas Anti-immigrant Group," *Intelligence Project*, January 25, 2005, http://www.splcenter.org/intel/news/item.jsp?aid=8 (accessed April 9, 2007); "Virginia Abernethy," *One Peoples Project*, February 28, 2004, http://www.onepeoplesproject.com/index.php?option=content&task=view&id=4 (accessed April 9, 2007). For an article by Virginia Abernethy, see "Immigration Moratorium: Effective Path to a Living Wage for Working Americans," *Population-Environment Balance*, 1999, http://www.balance.org/articles/immwages.html (accessed April 9, 2007).

20. On October 11, 2006, Pimentel issued a press release stating: "I strongly endorse Dr. Virginia Abernethy's [*sic*] of the Census Bureau Distortions that hide the serious immigration crisis that impacts on the people living in the United States. My concerns center on the effect of a rapidly escalating population has on all our basic natural resources that sustain life for everyone. . . . Each person uses fossil energy in the equivalent 2,800 gallons of oil per year with 500 gallons devoted for food. . . . Illegal immigrants are stressing U.S. schools and medical facilities, increasing disease incidence (like tuberculosis), causing unemployment, and causing a drain on the economy." Available online at http://www.carryingcapacity.org/pimentel.html (accessed November 20, 2006).

21. F. A. Hayek was an Austrian émigré economist whose book *The Road to Serfdom* (Chicago: University of Chicago Press, 1994) is the Bible of modern libertarian economic and social thought. This very powerful work was originally published in 1944, at the height of World War II, and it gained both acclaim and notoriety by its forceful arguments against socialism. It is based on its teachings, degraded into dogma, that many self-described libertarians (or classical liberals) oppose "on principle" a government mandate for flex-fuel vehicles. They would do well to read the book again, as Hayek himself is very clear about the need for government action to protect the free

market against monopolies. For example, see p. 21: "There is, in particular, all the difference between deliberately creating a system within which competition will work as beneficially as possible and passively accepting institutions as they are. Perhaps nothing has done so much harm to the liberal cause as the wooden insistence of some liberals on certain rough rules of thumb, above all the principle of laissez faire." Or p. 102: "Our freedom of choice in a competitive society rests on the fact that, if one person refuses to satisfy our wishes, we can turn to another. But if we face a monopolist we are at his mercy." Or p. 218: "[M]ake the position of the monopolist once more that of the whipping boy of economic policy. . . ." I submit that Hayek would have had no problem with the idea of using government action to break the oil cartel's vertical monopoly on the world's fuel supply.

CHAPTER 8

1. Shaohua Chen and Martin Ravallion, "How Have the World's Poorest Fared since the Early 1980s," Development Research Group, World Bank Policy working paper no. 3341, June 2004.

2. Borgna Brunner, ed., *Time Almanac 2006* (Boston, MA: Information Please, 2005). Statistics cited are based on data collected for 2004.

3. William Easterly, *The White Man's Burden: Why the West's Efforts to Aid the Rest Have Done So Much Ill and So Little Good* (New York: Penguin, 2006).

4. Jeffrey Sachs, *The End of Poverty: Economic Possibilities for Our Time* (New York: Penguin, 2005).

5. Easterly, *The White Man's Burden*, pp. 144–45.

6. Ibid., p. 133.

7. Jennifer Lake and Ralph Chite, "Emergency Supplemental Appropriations for Hurricane Katrina Relief," CRS Report for Congress, September 9, 2005, http://www.opencrs.com/getfile.php?rid=50233 (accessed April 4, 2007).

8. Easterly, *The White Man's Burden*, p. 342.

9. Joseph Stiglitz and Andrew Charlton, *Fair Trade for All* (Oxford: Oxford University Press, 2005), p. 42.

10. Ibid., p. 47.

11. Ibid., p. 57.

12. Data from *CIA World Fact Book*, available online at http://en.allexperts.com/e/l/li/list_of_countries_by_agricultural_output.htm (accessed April 4, 2007). Note that I have consolidated all EU nations into one entry.

13. In table 8.2, yields for all crops grown in the United States are taken from US Department of Agriculture figures, available online at http://usda.mannlib.cornell.edu/usda/current/CropProd/CropProd-03–09–2007.txt (accessed April 7, 2007). The numbers for ethanol yields for sugarcane and corn are current industry values. For barley, oats, rice, wheat, and sorghum, I have used ethanol yield numbers provided in Seungdo Kim and Bruce Dale, "Global Potential Bioethanol Production from Wasted Crops and Crop Residues," *Biomass and Bioenergy* 26 (2004): 361–75, http://www.icpf.cas.cz/trogl/download/bioethanol/uroda.pdf (accessed April 3, 2007). For the rest, I have used the somewhat dated National Research Council, Board on Science and Technology for International Development, Office of International Affairs, *Alcohol Fuels: Options for Developing Countries*, report of the Ad Hoc Panel of the Advisory Committee on Technology Innovation (Washington, DC: National Academy Press, 1983). These older NRC estimates may be conservative, since crop yields have improved significantly since that time.

14. Lester Brown, *Plan B 2.0: Rescuing a Planet under Stress and a Civilization in Trouble* (New York: W. W. Norton, 2006).

15. Lester Brown, "Mixed Blessings: Can Biofuel Change the World? Yes—For Better and for Worse," *World Ark*, September–October 2006.

16. Current sweet potato yields in Israel are 80 tonnes per hectare per year, or triple the US average cited here. See http://www.hort.purdue.edu/newcrop/duke_energy/Ipomoea_batatas.html (accessed April 7, 2007).

17. National Research Council, *Alcohol Fuels*.

18. Brown, "Mixed Blessings."

19. Data from F. O. Licht, Worldwatch, available online at http://www.earth-policy.org/Updates/2006/Update55_data.htm (accessed April 4, 2007).

20. V. Smil, "Crop Residues: Agriculture's Largest Harvest," *Bioscience* 49, no. 4 (1999): 299–308.

21. George Olah, Alain Goeppert, and G. K. Surya Prakash, *The Methanol Economy* (Weinheim, Germany: Wiley-VCH, 2006), p. 234.

22. American Methanol Institute, *The Promise of Methanol Fuel Cell Vehicles* (Washington, DC: AMI, 2000).

23. Olah, Goeppert, and Prakash, *The Methanol Economy*, chap. 13.

24. DME can also be mixed with either conventional diesel fuel or biodiesel. In such mixtures, the vapor pressure exerted by DME can be brought down roughly in proportion to its fraction of the fuel, so, for example, if it is used as a 20 percent component, the vapor pressure of the fuel would only be about 15 psi, instead of the 75 psi exerted by pure DME. DME-biodiesel mixtures are especially interesting, because pure biodiesel has too high a viscosity and freezing point for many (especially cold-weather) applications, and the addition of DME can solve both of those problems, while raising the fuel's cetane rating as well.

CHAPTER 9

1. By 1888 slavery was already gradually declining in Brazil. Emperor Pedro II had declared the slave trade illegal in 1852, and ruled in 1871 that all children born henceforth to slaves be free. Nevertheless, Princess Isabel's bold seizure of her moment to end slavery immediately with a stroke of the pen has to be seen as inspired.

2. Alfred Szwarc, "Use of Bio-Fuels in Brazil," presentation to In-Session Workshop on Mitigation, SBSTA 21 / COP 10, Buenos Aires, December 9, 2004. Available online at http://unfccc.int/files/meetings/cop_10/in_session_workshops/mitigation/application/pdf/041209szwarc-usebiofuels_in_brazil.pdf (accessed April 6, 2007).

3. David Sandalow, "Ethanol: Lessons from Brazil," Brookings Institute white paper, 2006, http://www.brookings.edu/printme.wbs?page=/pagedefs/7e3965440e9bff40800093c40a1415cb.xml (accessed April 6, 2007).

4. EIA data, available online at http://www.eia.doe.gov/emeu/cabs/topworldtables3_4.html and http://www.eia.doe.gov/emeu/cabs/topworldtables1_2.html (accessed April 3, 2007).

5. José Goldemberg and Suani Coelo, "Why Alcohol Fuel: The Brazilian Experience," CTI Industry Seminar on Technology Diffusion of Energy Efficiency in Asian Countries, Beijing, February 24–25, 2005, http://www.resourcesaver.com/file/toolmanager/O105UF1261.pdf#search=percent22Josepercent20Goldembergpercent22 (accessed October 3, 2006).

6. C. Almeida, M. Kotek, M. Pereira, and P. Albuquerque, "Alternative

Fuel in Brazil: Flex-Fuel Vehicles," white paper prepared for University of North Carolina MBA program, 2004, http://www.cse.unc.edu (accessed October 3, 2006).

7. For information on the growing Chinese ethanol program, see "China Promotes Ethanol-Based Fuel in Five Cities," *People's Daily*, June 17, 2002, http://english.people.com.cn/200206/17/eng20020617_98009.shtml (accessed April 5, 2007); "China to Fill Cars with Cassava Ethanol," Reuters, June 19, 2006, http://www.planetark.org/dailynewsstory.cfm/newsid/36881/story.htm (accessed April 5, 2007); "China's Medium to Long Term Energy Development Plan," February 2007, http://eneken.ieej.or.jp/en/data/pdf/383.pdf.

8. "India Rolls Out E5," *Biofuel Review*, www.biofuelreview.com/content/view/315/2/ (accessed October 5, 2006).

9. John Mathews, "A Biofuels Manifesto: Why Biofuels Industry Creation Should Be 'Priority Number One' for the World Bank and for Developing Countries," Macquarie University, Graduate School of Management, October 2006, http://www.gsm.mq.edu.au/facultyhome/john.mathews/a%20Biofuels%20manifesto%20%209%20oct%2006.pdf (accessed April 5, 2007).

10. Since these lines were written, Senator Barack Obama (D-IL) has emerged as a significant challenger for the Democratic presidential nomination. In 2006 Obama was one of only four senators to cosponsor a bill calling for a flex-fuel mandate for all US automobiles, albeit phased in over a ten-year time scale (see chap. 13). This speaks well for him. However, in his book, *The Audacity of Hope* (New York: Crown, 2006), Obama calls for an energy policy of soaking the oil companies with taxes and using the proceeds to pay for a random assortment of government-funded energy-related research and development projects, making no mention of a flex-fuel mandate. Such a policy would lead nowhere. Hopefully, Obama will improve his energy program as his candidacy matures. Former New York mayor Rudolf Giuliani has also said that he favors more ethanol development, but has given no specifics.

11. Quoted in Yossef Bodansky, *Bin Laden, the Man Who Declared War on America,* (Rocklin, CA: Forum Books, 1999), p. 322.

12. Rachel Ehrenfeld, *Funding Evil: How Terrorism Is Financed—and How to Stop It* (Chicago: Bonus Books, 2003).

13. Ibid., p. 56.

14. Germano Oliveira, "Brazil's Former Drug Czar: Bin Laden Estab-

lishing Al Qaeda Cell on Triborder," *O Globo*, September 11, 2001; trans. Foreign Broadcast Information Service, September 19, 2001.

15. Jerry Speer, "Terror Cell on Rise in South America," *Washington Times,* December 18, 2002.

16. Ehrenfeld, *Funding Evil*, p. 150.

17. Fabio Castillo, with research by Leydi Herrera, "The Hezbollah Contact in Colombia," part 3 of "Tracking the Tentacles of the Middle East in South America," *El Espectador*, December 9, 2001; trans. Foreign Broadcast Information Service, December 10, 2001.

18. Ehrenfeld, *Funding Evil*, p. 146. See also LaVerle Berry et al., "A Global Overview of Narcotics-Funded Terrorist and Other Extremist Groups," Federal Research Division, Library of Congress, May 2002, http://209.85.165.104/search?q=cache:S0PnJU29J48J:www.loc.gov/rr/frd/pdf-files/NarcsFundedTerrs_Extrems.pdf+A+Global+Overview+of+Narcotics-Funded+Terrorist+and+other+Extremist+Groups&hl=en&ct=clnk&cd=1&gl=us&ie=UTF-8 (accessed October 11, 2006).

19. Office of the United States Trade Representative, "First Report to Congress on the Andean Trade Preference Act, as Amended," April 2003.

20. Laura Chasen Cohen, international trade consultant, private communication, September 16, 2006.

21. Ehrenfeld, *Funding Evil*, p. 207.

22. Jim Robbins, "Drug War Awaits Attack of Killer Fungus," *New York Times*, July 18, 2000. The article also discusses the work of Montana State University Professor David Sands, who is one of the leading researchers in the area of mycoherbicides.

23. US Department of State, "Fact Sheet on Microherbicide Cooperation," July 17, 2000, http://www.ciponline.org/colombia/071801.htm (accessed October 12, 2006).

CHAPTER 10

1. Based on the 1/4 power law, Venus's absolute temperature should be 1.19 times as great as Earth's. Multiplying this factor by a typical Earth temperature of 283 K (10°C or 50°F), we get an absolute temperature of 337 K for Venus, which is equal to 64°C or 147°F.

2. The Impact Team, *The Weather Conspiracy: The Coming of the New Ice Age* (New York: Ballantine Books, 1977).

3. Stephen Schneider and Randi Londer, *The Coevolution of Climate and Life* (San Francisco, CA: Sierra Club Books, 1984).

4. Patrick J. Michaels, *Meltdown: The Predictable Distortion of Global Warming by Scientists, Politicians, and the Media* (Washington, DC: Cato Institute, 2004).

5. Al Gore, *An Inconvenient Truth: The Planetary Emergency of Global Warming and What We Can Do about It* (New York: Rodale Books, 2006).

6. The Hadley Centre for Climate Prediction and Research of the UK Meteorological Office. Graph courtesy of Global Warming Art, http://www.globalwarmingart.com/wiki/Image:Instrumental_Temperature_Record_png (accessed April 6, 2007).

7. Tidal gauge data from Bruce C. Douglas, "Global Sea Rise: A Redetermination," *Surveys in Geophysics* 18 (1997): 279–92. Satellite data from NASA's TOPEX/Poseidon oceanographic satellite. Graph courtesy of Global Warming Art, http://www.globalwarmingart.com/wiki/Image:Recent_Sea_Level_Rise_png (accessed April 7, 2007).

8. C. D. Keeling and T. P. Whorf, "Atmospheric Carbon Dioxide Record from Mauna Loa," Carbon Dioxide Group, Scripps Institute of Oceanography, http://cdiac.esd.ornl.gov/trends/co2/sio-mlo.htm (accessed October 16, 2006). Graph courtesy of Global Warming Art, http://www.globalwarmingart.com/wiki/Image:Mauna_Loa_Carbon_Dioxide_png (accessed April 6, 2007).

9. James Kasting, "The Carbon Cycle, Climate, and the Long-Term Effects of Fossil Fuel Burning," *Consequences* 4, no. 1 (1998), http://www.gcrio.org/CONSEQUENCES/vol4no1/carbcycle.html (accessed October 16, 2006).

10. As can be seen from examining table 5.7, humanity's current global carbon fuel usage is roughly 10 Gt per year and about 90 percent of that, or 9 Gt per year, is burned as fuel. However, in 1958 it was only about 2 Gt, with the average over the fifty-year period being about 5 Gt. Thus, in the past half century we have cumulatively released about 250 Gt of carbon into the atmosphere.

11. A. B. Robinson, S. L. Baliunas, W. Soon, and Z. W. Robinson, "Environmental Effects of Increased Atmospheric Carbon Dioxide," *Climate Research* 13 (1999): 149–64, http://www-pord.ucsd.edu/~sgille/stpa35/petition_justification.pdf (accessed April 6, 2007).

12. Oxygen comes naturally in two isotopes, the common "light" isotope O-16, with an atomic weight of 16, and the rarer "heavy" isotope O-18, with an atomic weight of 18. During periods of cold climate, such as an ice age, increased amounts of O-16 are trapped preferentially in ice sheets near the poles. This leaves the ocean waters comparatively enriched in O-18, and the record of such past enrichment will be preserved in marine fossils laid down at that time. Thus, by measuring the ratio of O-18/O-16 in marine fossils, scientists can assess global temperatures in ages past. The higher the O-18/O-16 ratio, the colder the global climate. For more on this technique, see NASA Earth Observatory, "Paleoclimatology: The Oxygen Balance," http://earthobservatory.nasa.gov/Study/Paleoclimatology_OxygenBalance/oxygen_balance.html (accessed April 5, 2007).

13. Schneider and Londer, *The Coevolution of Climate and Life*, p. 111.

14. R. S. Bradley and J. A. Eddy, *EarthQuest*, vol. 1, 1991; Based on J. T. Houghton et al., *Climate Change: The IPCC Assessment* (Cambridge: Cambridge University Press, 1990), http://gcrio.org/CONSEQUENCES/winter96/article1-fig1.html (accessed April 9, 2006).

15. James Zachos, Mark Pagani, Lisa Sloan, Ellen Thomas, and Katharina Billups, "Trends, Rhythms, and Aberrations in Global Climate 65 Ma to Present," *Science* 292, no. 5517 (2001): 686–93, http://www.DOI:10.1126/science.1059412 (accessed October 21, 2006). Graph courtesy of Global Warming Art, http://www.globalwarmingart.com/wiki/Image:65_Myr_Climate_Change_Rev_png (accessed April 6, 2007).

16. Tim K. Lowenstein and Robert V. Demicco, "Elevated Eocene Atmospheric CO_2 and Its Subsequent Decline," *Science* 313, no. 5795 (2006): 1928. The authors give estimates of early Eocene CO_2 levels ranging from 1,125 ppm to 3,000 ppm. I have simplified this to ~2,000 ppm.

17. Bonnie F. Jacobs, John D. Kingston, and Louis L. Jacobs, "The Origin of Grass-Dominated Ecosystems," *Annals of the Missouri Botanical Garden* 86, no. 2 (1999): 590–643.

18. Louis D. Johnson, "What Was GDP Then?" http://eh.net/hmit/gdp/.

19. David F. Nygard, "World Population Projections 2020," International Food Policy Research Institute, October 1994, http://www.ifpri.org/2020/briefs/number05.htm (accessed April 5, 2007).

20. "UM Study Shows Increased Plant Growth," University of Montana press release, 2002, http://www.umt.edu/urelations/vision/2002/9plant.htm (accessed October 22, 2006).

21. "Carbon Dioxide Fertilization Is Neither Boon nor Bust," Oak Ridge National Lab press release, February 15, 2004, http://www.eurekalert.org/pub_releases/2004–02/jaaj-cdf020504.php (accessed October 22, 2006).

CHAPTER 11

1. George Olah, Alain Goeppert, and G. K. Surya Prakash, *The Methanol Economy* (Weinheim, Germany: Wiley-VCH, 2006), pp. 3, 100, 121. See also EIA statistics at http://www.eia.doe.gov/fuelelectric.html (accessed April 5, 2007).

2. William Sweet, *Kicking the Carbon Habit: Global Warming and the Case for Renewable and Nuclear Energy* (New York: Columbia University Press, 2006).

3. H. Keith Florig, "China's Air Pollution Risks," *Environmental Science and Technology* 32, no. 6 (1997): 274–79.

4. World Coal Institute, "Clean Coal: Building a Future through Technology," 2004, http://www.worldcoal.org/pages/content/index.asp?PageID=36 (accessed October 24, 2006).

5. Bjorn Lomberg, *The Skeptical Environmentalist* (Cambridge: Cambridge University Press, 2001), p. 285.

6. The Chernobyl Forum, *Chernobyl's Legacy: Health, Environmental, and Socioeconomic Impacts* (Vienna: IAEA, September 2005). A twenty-year assessment of the consequences of the Chernobyl accident by the IAEA, WHO, and FAO.

7. Charles D. Hollister, D. Richard Anderson, G. Ross Heath, "Subseabed Disposal of Nuclear Wastes," *Science* 213, no. 4514 (1981): 1321–26. See also "Seabed Nuclear Waste Solutions," *Scientia Press*, http://www.scientiapress.com/findings/sea-based.htm (accessed April 5, 2007).

8. This is because commercial reactors keep their fuel in place for a long time, during which some of the Pu-239 created in the reactor absorbs a further neutron to become Pu-240. The Pu-240 seriously degrades the value of the plutonium for weapons purposes. However, in stand-alone atomic piles, such as those developed in Hanford during the World War II Manhattan Project, the fuel is not left in the system for long, so the plutonium produced is not spoiled.

9. Ronald Knief, *Nuclear Energy Technology* (New York: McGraw-Hill, 1981), p. 549. See also Olah et al., *The Methanol Economy*, pp. 27–50.

10. R. W. Conn et al., "Lower Activation Materials and Magnetic Fusion Reactors," *Nuclear Technology/Fusion* 5, no. 291 (1984); G. R. Hopkins et al., "Low Activation Fusion Reactor Design Studies," Fifth Topical Meeting on Technology of Fusion Energy, Knoxville, TN, April 26–28, 1983, abstract available online at http://adsabs.harvard.edu/abs/1983tfe.meetS.26H (accessed April 5, 2007).

11. R. W. Conn, "Magnetic Fusion Reactors," in *Fusion*, vol. 1: *Magnetic Confinement*, part B, ed. E. Teller (New York: Academic Press, 1981).

12. R. Zubrin, "A Deuterium-Tritium Ignition Ramp for an Advanced Fuel Field-Reversed Configuration Reactor," *Fusion Technology* 9, no. 1 (1986): 97–100.

13. In addition to leading the development of the Soviet hydrogen bomb, Sakharov was the inventor of the tokamak. Artsimovich became the concept's virtuoso.

14. The idea of employing a scaled-up magnetic field, rather than an enlarged machine, as a way of obtaining fusion ignition conditions cheaply was first proposed by visionary scientist Robert Bussard and maverick MIT physicist Bruno Coppi in the 1980s. Bussard attempted to raise private money to build such a device, but was unsuccessful. Coppi, who called his version of the concept an "Ignitor," tried pushing it within the system, but was unable to make the idea prevail at that time, so it was not incorporated into ITER. Over the years since, however, it has become more accepted, and the Ignitor-descended FIRE concept is now championed by such mainstream fusion program leaders as Dale Meade, the director of the Princeton Plasma Physics Lab. The fact that they can more easily adapt to take advantage of new creative ideas is another reason to prefer comparatively compact national fusion programs to grand international consortia.

15. Deuterium is five times as common on Mars as it is on Earth, and thus could provide a plentiful energy source to space pioneers possessing fusion technology. Fusion rockets could theoretically produce exhaust velocities of 8 percent the speed of light. Since rockets can be engineered to achieve about twice their exhaust velocity, such systems could make interstellar travel within a human lifetime possible. For more on such possibilities, see R. Zubrin with R. Wagner, *The Case for Mars: The Plan to Settle the Red Planet and Why We Must* (New York: Free Press, 1996); and R. Zubrin,

Entering Space: Creating a Spacefaring Civilization (New York: Tarcher Putnam, 1999).

CHAPTER 12

1. Winston Churchill, *The World Crisis, 1911–1918* (New York: Charles Scribner's Sons, 1931; reprint, New York: Free Press, 2005), pp. 75–76.

2. In the original French, Gallieni's famous remark was "Eh bien, voilà au moins qui n'est pas banal!"

3. Daniel Yergin, *The Prize: The Epic Quest for Oil, Money, and Power* (New York: Touchstone Books, 1991), p. 171. This book is the single best source for the history of the role of oil in world affairs.

4. In addition to extending a ship's range, the use of oil also allowed a reduction in the number of crew members needed as stokers, with correspondingly more sailors available to man the guns. Oil-powered ships were faster, too, and for Fisher and Churchill, this was decisive. At one prewar parliamentary debate, Churchill was criticized for preferring to build fast oil-powered destroyers instead of slower but cheaper coal-fired ones. Churchill famously replied, "Build slow destroyers! One might just as well breed slow race horses."

5. Yergin, *The Prize,* pp. 176–77.

6. Ibid., p. 183.

7. Ibid., pp. 328–31.

8. There are innumerable accounts available of the battle of Stalingrad, but in my view the best by far is that presented by Vasily Grossman in his novel *Life and Fate*, trans. Robert Chandler (New York: Harper & Row, 1985). Grossman was a dissident Soviet Jewish physicist, but during World War II he served as a combat correspondent, covering the battle of Stalingrad for the army newspaper *Red Star*. After the war, he wrote up his experience in the form of this phenomenal Tolstoyan epic novel. There are scenes in this book you will not forget for as long as you live. In the opinion of many, it is the greatest work of twentieth-century Russian literature. The book was banned by the Soviet authorities because, despite its patriotic theme, it presented far too penetrating an analysis of the nature of the totalitarian mind.

Remarkable reportage of Stalingrad can also be found in British journalist Alexander Werth's superb *Russia at War: 1941–1945* (New York: Avon, 1964). He was there, too.

9. Yergin, *The Prize*, p. 337.

10. Richard Overy, *Why the Allies Won* (New York: W. W. Norton, 1995), chap. 2. It was the application of air power at sea that defeated the U-boats. The key step was taken in April and May 1943, when, at the insistence of Admiral Max Horton, escort carriers mounting continuous air patrols were finally deployed with all convoys. In March 1943, eighty-two Allied ships totaling 700,000 tons were sunk by U-boats. In April, shipping losses were cut in half. By May, Allied losses were down to 160,000 tons, while forty-one U-boats were sunk. In June 1943, Allied shipping losses to U-boats were zero. The U-boats never found an adequate response. Overall, in 1944, Allied losses to U-boats were only 3 percent of what they were in 1942.

11. Richard G. Davis, "General Carl Spaatz and D-Day," *Airpower Journal*, Winter 1997, http://www.airpower.maxwell.af.mil/airchronicles/apj/apj97/win97/davis.html (accessed April 5, 2007).

12. The Ploesti raid of August 1, 1943, is one of the epic episodes in the history of the US Air Force. Five of those participating won the Congressional Medal of Honor. For an overview of the raid, see Robin Neillands, *The Bomber War* (New York: Barnes and Noble, 2001), pp. 244–47. For a blow-by-blow collection of accounts, see "Ploesti: When Heroes Filled the Sky," Home of Heroes, http://www.homeofheroes.com/wings/part2/09_ploesti.html (accessed April 5, 2007).

13. Albert Speer, *Inside the Third Reich* (New York: Avon Books, 1970), especially chap. 24, "The War Thrice Lost," and chap. 27, "The Wind from the West."

14. Overy, *Why the Allies Won*, pp. 331–32.

15. The *Liebstandarte Adolf Hitler* was Peiper's unit's name. It means "Adolf Hitler's bodyguard." An Obersturmbannführer was the SS equivalent of an army colonel. Peiper earned his posting at the head of this elite Nazi formation through his earlier role as commander of the SS's "Blowtorch Battalion" on the eastern front in 1943, where he specialized in herding Russian and Ukrainian villagers into their churches and setting them afire.

16. For a great account of the action where the American 291st engineering battalion stopped Peiper's 1st SS Panzer division at Stavelot, see Janice Holt Giles, *The Damned Engineers* (New York: Houghton Mifflin,

1970). Giles, an accomplished writer, was the wife of Sergeant Henry Giles of the 291st. She interviewed all the survivors of the unit after the war, and gave the GIs a voice. Peiper was captured and sentenced to death for his role in the Malmedy massacre. His sentence, however, was commuted to life imprisonment, and he was released from jail in 1956. After evading prosecution for various other atrocities, he took up residence in France, where the highly educated Nietzschean found comfortable employment as a literary translator. This situation was rectified in 1976, when unknown assailants, no doubt still irate over some aspect of Peiper's past conduct, broke into his house and burned him alive. Peiper remains a cult hero today to neo-Nazis. In their literature, they claim he was done in by "French Communists." That might be true. I, however, prefer to believe the work of sending the former Obersturmbannführer to his infernal reward was accomplished by middle-aged American tourists—you know, some of those fifty-something characters who used to hang around the VFW posts in large numbers back in the seventies telling war stories to anyone who would listen. Well done, guys.

17. Gordon Prange, *Pearl Harbor: The Verdict of History* (New York: McGraw-Hill, 1981), p. 566.

18. Other forces also contributed to the destruction of Japanese merchant shipping, but I emphasize the US submarines because with just 2 percent of naval manpower, they were responsible for 55 percent of all sinkings. By 1945, 86 percent of Japan's merchant fleet would be sunk, and another 9 percent crippled.

19. Yergin, *The Prize*, p. 361.

20. Masanori Ito, *The End of the Imperial Japanese Navy* (New York: McFadden Books, 1965), p. 129. Ito notwithstanding, the destroyer action at Leyte Gulf was the US Navy's finest hour. For American accounts of the battle spanning three generations of historians, see C. Vann Woodward, *The Battle for Leyte Gulf* (New York: Ballantine Books, 1946); Edwin P. Hoyt, *The Battle of Leyte Gulf* (New York: Jove Books, 1972); and James D. Hornfischer, *The Last Stand of the Tin Can Sailors* (New York: Bantam Books, 2004).

CHAPTER 13

1. As of April 9, 2007, the DRIVE Act has eighty-two sponsors in the House and twenty-six sponsors in the Senate, notably including senators Hillary Clinton (D-NY), Barack Obama (D-IL), and Sam Brownback (R-KS). The text of the bill and a complete list of sponsors can be found at http://www.setamerica free.org/pc0118.htm (accessed April 9, 2007).

2. The characteristic Christian view of God as "Our Father" is a development that originates in Judaism. "Father" is used as a metaphor in Psalm 68:5 and Psalm 103:13. God is directly addressed as "Father" in Psalm 89:26: "He (David) shall cry to me 'You are my Father, My God, and the Rock of my salvation.'" In the Prophets, the description of God as Father becomes more frequent. For example, we have from Isaiah 9:6: "Wonderful Counselor, Mighty God, Everlasting Father, Prince of Peace . . ." (famously incorporated into Handel's *Messiah*), as well as Isaiah 63:16: "For Thou art our father . . ." and ". . . you, O Lord, are our Father" and Isaiah 64:8: "Yet, O lord, you are our Father, . . ." References to God as Father also appear in Malachi and Jeremiah.

3. The denial of cause and effect in Islam shocked Western rationalist sensibilities as early as the twelfth century. For example, in a gloss written circa 1141 titled *Summary of the Whole Heresy of the Diabolical Sect of the Saracens*, a Cluniac monk offers the following critique of the Koran: "See the simplicity of this madman [Mohammad] who thinks that flying birds are supported not by the air but by the miraculous power of God. But as we know, fishes are supported by water and birds by air, according to an original decree and ordering of God, and not, as he thinks, by a special and invisible miracle" (quoted from R. W. Southern, *The Making of the Middle Ages* [New Haven, CT: Yale University Press, 1953]). In this one quote, a civilization-defining difference between Christianity and Islam is made evident. While both have a theistic outlook, the Cluniac's view allows for the development of science, whereas Mohammad's does not.

4. Bat Ye'or, *Islam and Dhimmitude: Where Civilizations Collide* (Cranbury, NY: Fairleigh Dickinson University Press, 2002), pp. 44, 59.

5. The English version of the Koran officially endorsed by the Saudi Arabian government is the Tahrike Tarsile Qur'an, translated by Abdullah Yusuf Ali. The sixteenth paperback edition cited here is published by Tahrike Tarsile Qur'an, Elmhurst, NY: 2006, p. 116.

6. Robert Spencer, *Islam Unveiled* (San Francisco: Encounter Books, 2002).

7. Bat Ye'or, *Islam and Dhimmitude*, p. 60.

8. Peter Balakian, *The Burning Tigris: The Armenian Genocide and America's Response* (New York: HarperCollins, 2003).

9. Many of the leading thinkers of the Islamic golden age were actually not Muslims. The early philosopher Al Kindi, for example, was a Christian. Ibn Gabirol (aka Avicebron) was a Jew. The progressive degradation of the dhimmi population is viewed by a number of historians as an important contributing factor in the decline of Islamic culture.

10. Toby Huff, *The Rise of Early Modern Science: Islam, China, and the West* (Cambridge: Cambridge University Press, 1993).

11. Ibn Warraq, *Why I Am Not a Muslim* (Amherst, NY: Prometheus Books, 1995).

12. Ibn Warraq, a former Muslim, describes the sexual practices that polygamy forces upon the majority of males such a system necessarily leaves without women. I will forgo the lurid details. Those interested can find them in *Why I Am Not a Muslim*, pp. 340–43.

13. Al Gore, *Earth in the Balance* (New York: Plume Books, 1993), p. 269.

POSTSCRIPT: IN DEFENSE OF BIOFUELS

1. Muriel Boselli, "Saudi Oil Minister Slams Biofuels, Supports Solar Energy," Reuters, April 10, 2008, http://uk.reuters.com/article/environment News/idUKL1079284820080410 (accessed September 6, 2008).

2. "OPEC President Blames Ethanol for Crude-Price Rise," *Market-Watch*, July 6, 2008, http://www.marketwatch.com/news/story/opec -president-blames-oil-prices/story.aspx?guid=%7BE003D4C9-0739-4868 -8F69-D9C51BB53CB5%7D (accessed September 6, 2008).

3. "Chavez Calls Ethanol Production 'Crime,'" Associated Press, April 26, 2008, http://www.newsmax.com/international/venezuela_ethanol/2008/ 04/26/91282.html (accessed September 6, 2008).

4. "Report of the Attorney General to the Congress of the United States on the Administration of the Foreign Agents Registration Act of 1938, as

Amended, for the First Six Months Ending June 30, 2007," US Department of Justice, Washington DC, www.usdoj.gov/criminal/fara/reports/June30-2007.pdf (accessed September 6, 2008). Glover Park was also retained by the government of Dubai to assist its effort to gain control on America's ports. See Dick Morris and Eileen McGann, "Hill and Lobbyists: The More the Merrier," *New York Post*, April 9, 2008, http://www.nypost.com/seven/04092008/postopinion/opedcolumnists/hill__lobbyists__more_the_merrier_105664.htm (accessed September 8, 2008).

5. Anna Palmer, "Beating Up on Ethanol," *Roll Call*, May 14, 2008, http://www.rollcall.com/issues/53_137/news/23620-1.html?type=printer_friendly (accessed September 6, 2008). Palmer's article included documents showing that the Grocery Manufacturers Association, represented by one Scott Faber, a former staff member with Searchinger's Environmental Defense, had paid Glover Park $300,000 to launch an "aggressive public relations campaign" to "obliterate whatever intellectual justification might still exist for corn-based ethanol among policy elites." There are also reports of involvement in the antiethanol campaign by the Daniel J. Edelman Communications firm, whose clients include the American Petroleum Institute, the Grocery Manufacturers Association, and Environmental Defense. As of this writing, the reports of Edelman's involvement have not been documented. However, the remarkable confluence of its client list with the antiethanol campaign principals makes the matter worthy of further investigation.

6. Connie Hardy, "Ethanol Profile," Agricultural Marketing Research Center, http://www.agmrc.org/agmrc/commodity/grainsoilseeds/corn/ethanolprofile.htm (accessed September 9, 2008).

7. Aditya Chakrabortty, "Secret Report: Biofuel Caused Food Crisis," *Guardian*, July 4, 2008, http://www.guardian.co.uk/environment/2008/jul/03/biofuels.renewableenergy (accessed September 9, 2008).

8. Grant Ferret, "Biofuels 'Crime against Humanity,'" BBC News, October 27, 2007, http://news.bbc.co.uk/2/hi/americas/7065061.stm (accessed September 8, 2008).

9. Jeffrey Sachs, "Surging Food Prices Mean Global Instability," *Scientific American*, May 19, 2008, http://www.earth.columbia.edu/articles/view/2198 (accessed September 9, 2008).

10. US Department of Agriculture, Economic Research Service, http://www.ers.usda.gov (accessed September 8, 2008).

11. "Outlook for Agricultural Trade," AES-59 US Department of Agri-

culture, August 28, 2008, http://usda.mannlib.cornell.edu/MannUsda/view DocumentInfo.do?documentID=1196 (accessed September 8, 2008).

12. Gail L. Cramer and Clarence Jensen, *Agricultural Economics and Agribusiness*, 6th ed. (New York: John Wiley and Sons, 1994).

13. "Ethanol's Impact on Food and Prices," National Corn Growers Association, http://www.ncga.com/news/publications/PDF/GetTheFactsOn FoodAndFuel.pdf (accessed September 8, 2008).

14. Patrick Barta, "World's Reliance on Biofuels Grows," *Wall Street Journal*, March 25, 2008, http://www.theaustralian.news.com.au/story/ 0,25197,23424736-36375,00.html (accessed February 6, 2009).

15. Adam Smith, *The Wealth of Nations*, bk. 4 (New York: Modern Library, 1994).

16. Timothy Searchinger et al., "Use of U.S. Croplands for Biofuels Increases Greenhouse Gases through Emissions from Land-Use Change," *Science* 319 (February 29, 2008): 1238–40. The study was subsequently circulated in a more readable form authored by Searchinger alone as a policy paper issued by the German Marshall Fund, "The Impacts of Biofuels on Greenhouse Gases: How Land Use Change Alters the Equation," http://www .gmfus.org/publications/article.cfm?id=385 (accessed September 8, 2008).

17. Roger Bate and Richard Tren, *Malaria and the DDT Story* (London: Institute for Economic Affairs, 2001), available online at www.fighting malaria.org/pdfs/malaria_and_DDT_story_IEA.pdf (accessed February 6, 2009).

18. Michael Grunwald, "The Clean Energy Scam," *Time*, March 27, 2008, http://www.time.com/time/magazine/article/0,9171,1725975,00.html (accessed September 8, 2008). Grunwald was assisted in the writing of his article by Searchinger, who accompanied him on his trip to Brazil.

19. Foreign Agricultural Trade of the United States (FATUS), USDA Economic Research Service, "Data Sets," http://www.ers.usda.gov/Data/ FATUS/ (accessed February 6, 2009).

20. Michael Wang and Zia Haq, "Letter to *Science*," February 14, 2008, revised March 14, 2008, www.transportation.anl.gov/pdfs/letter_to_science _anldoe_03_14_08.pdf (accessed September 6, 2008).

21. Marco Sibaja, "Brazil Police Seize Amazon Lumber," Associated Press, February 14, 2008, http://www.ibtimes.com/articles/20080214/brazil -police-seize-amazon-lumber.htm (accessed September 8, 2008).

SELECTED BIBLIOGRAPHY

ON ISLAM, SAUDI ARABIA, AND TERRORISM

Aburish, Said K. *The Rise, Corruption, and Coming Fall of the House of Saud*. New York: St. Martin's Press, 1995.

Ali, Ayaan Hirsi. *Infidel*. New York: Free Press, 2007. A moving biography telling what it is like to be a woman living in an Islamic society.

Baer, Robert. *See No Evil: The True Story of a Ground Soldier in the CIA's War on Terrorism*. New York: Three Rivers Press, 2002. Contains much good anecdotal material relating to terrorist activity in the Middle East and central Asia.

———. *Sleeping with the Devil: How Washington Sold Our Souls for Saudi Crude*. New York: Three Rivers Press, 2003. A racy introduction into Saudi influence peddling among American elites, written by a former CIA agent.

Balakian, Peter. *The Burning Tigris: The Armenian Genocide and America's Response*. New York: HarperCollins, 2003.

Berlinski, Claire. *Menace in Europe: Why the Continent's Crisis Is America's Too*. New York: Crown Forum, 2006. This elegantly written book by an overseas American is one of an entire genre of works warning about the ongoing dangerous retreat of Western culture and values in Europe in the face of Islamist penetration. Interestingly, almost all such works seem to be written by women, who indeed have the most urgent cause for alarm.

Others of this type include the works by Oriana Fallaci, Melanie Phillips, and Bat Ye'or, cited below.

Bostom, Andrew G., ed. *The Legacy of Jihad: Islamic Holy War and the Fate of Non-Muslims.* Amherst, NY: Prometheus Books, 2005.

Ehrenfeld, Rachel. *Funding Evil: How Terrorism Is Financed—and How to Stop It.* Chicago: Bonus Books, 2003. A fascinating examination of terror financing, with special emphasis on the growing links between terror groups and criminal organizations, especially those involved in narcotics traffic. Written by a former Drug Enforcement Agency official.

Fallaci, Oriana. *The Force of Reason.* New York: Rizzoli International, 2006. Fallaci expands her warnings against capitulation to Islamism in Europe in a more deliberate fashion. Quite a read.

———. *The Rage and the Pride.* New York: Rizzoli International, 2002. The late Oriana Fallaci was a leftist Italian journalist, famous for reporting from the front lines of revolutionary causes in Vietnam, Chile, and elsewhere. After September 11, 2001, however, she took alarm about the growth of Islamist influence in Europe, and published this very emotional book warning of the danger. In response, she was excommunicated by her former comrades on the Left and ultimately banned from Italy.

Gabriel, Brigitte. *Because They Hate: A Survivor of Islamic Terror Warns America.* New York: St. Martin's Press, 2006. A firsthand account of what it is like to be a Christian living under Islamic rule.

Gold, Dore. *Hatred's Kingdom.* Washington, DC: Regnery, 2003. The best short presentation on the history and character of Saudi Arabia.

Huff, Toby. *The Rise of Early Modern Science: Islam, China, and the West.* Cambridge: Cambridge University Press, 1993. A scholarly work that shows how Islam aborted the development of science within its realm at the same time that Christianity initiated the development of learning and inquiry in Europe during the High Middle Ages. Very useful for understanding the deep philosophical differences between the two creeds.

Lewis, Bernard. *The Crisis of Islam: Holy War and Unholy Terror.* New York: Random House, 2003. In his first book written after 9/11, Lewis takes off the gloves.

———. *Islam and the West.* Oxford: Oxford University Press, 1993.

———.*The Middle East: A Brief History of the Last 2,000 Years.* New York: Touchstone Books, 1995. The historical magnum opus of America's leading Middle East scholar.

———. *What Went Wrong? Western Impact and Middle Eastern Response.* Oxford: Oxford University Press, 2002. How Islamic fundamentalism has stifled the development of the Middle East.

Murawiec, Laurent. *Princes of Darkness: The Saudi Assault on the West.* Lanham, MD: Rowman & Littlefield, 2003. A useful source for information about Saudi overseas operations, including influence peddling in Washington, DC.

Napoleoni, Loretta. *Terror Incorporated: Tracing the Dollars behind the Terror Networks.* New York: Seven Stories Press, 2005. A view from the Left.

Phillips, Melanie. *Londonistan.* New York: Encounter Books, 2006. Describes the degradation of British values and institutions in the face of growing domestic Islamist influence. Tory journalist Phillips ascribes her government's incredible appeasement policy to terminal political correctness. It might have been interesting to also look for a money trail.

Pipes, Daniel. *Militant Islam Reaches America.* New York: W. W. Norton, 2002.

Posner, Gerald. *Secrets of the Kingdom: The Inside Story of the Saudi-US Connection.* New York: Random House, 2005.

———. *Why America Slept: The Failure to Prevent 9/11.* New York: Ballantine Books, 2003.

The Qur'an. Translated by Abdullah Yusuf Ali. Elmhurst, NY: Tahrike Tarsile Qur'an, 2006. This is the official Saudi government–approved translation.

Rashid, Ahmed. *Taliban: Militant Islam, Oil, and Fundamentalism in Central Asia.* New Haven, CT: Yale University Press, 2001. Rashid is a very respected journalist covering central Asian affairs. If you want to understand how the Taliban was created, this is the book to read.

Spencer, Robert. *Islam Unveiled.* San Francisco: Encounter Books, 2002.

———. *The Myth of Islamic Tolerance: How Islamic Law Treats Non-Muslims.* Amherst, NY: Prometheus Books, 2005.

———. *Onward Muslim Soldiers: How Jihad Still Threatens America and the West.* Washington, DC: Regnery, 2003.

Trifkovic, Serge. *The Sword of the Prophet: Islam; History, Theology, and Impact on the World.* Boston, MA: Regina Orthodox Press, 2002. A view from the Right.

Unger, Craig. *House of Bush, House of Saud: The Secret Relationship between the World's Two Most Powerful Dynasties.* New York: Scribner, 2004. A partisan treatment that limits its exposure of Saudi corruption of

Washington to Republican targets. Nevertheless, it contains some good data.

Warraq, Ibn. *Why I Am Not a Muslim.* Amherst, NY: Prometheus Books, 1995. A very powerful and in-depth treatment of Islam by a former believer.

Ye'or, Bat. *Eurabia: The Euro-Arab Axis*. Cranbury, NY: Fairleigh Dickinson University Press, 2005. A frightening discussion of the subversion of European political elites and institutions by Arab oil money.

———. *Islam and Dhimmitude: Where Civilizations Collide*. Cranbury, NY: Fairleigh Dickinson University Press, 2002. Ye'or is a Jew who grew up in Egypt. She knows about dhimmitude through both extensive scholarship and direct experience.

Zubrin, Robert. *The Holy Land.* Lakewood, CO: Polaris Books, 2003. The Middle East situation and international terrorism, dissected and examined through the instrument of science-fiction satire. In humor there is truth.

ON THE OIL INDUSTRY, ALCOHOL FUELS, AND ENERGY RESOURCES

American Methanol Institute. *The Promise of Methanol Fuel Cell Vehicles.* Washington, DC: AMI, 2000.

Brown, Charles. *World Energy Resources.* Berlin: Springer-Verlag, 2002.

Brunner, Borgna, ed. *Time Almanac 2006*. Boston, MA: Information Please, 2006.

———. *Time Almanac 2007*. Boston, MA: Information Please, 2006. A convenient resource for summary figures on energy production and consumption worldwide.

Doxon, Lynn Ellen. *The Alcohol Fuel Handbook.* Haverford, PA: Infinity Publishing, 2001. How to make ethanol from a variety of crops.

Economides, Michael, and Ronald Oligney. *The Color of Oil: The History, the Money and the Politics of the World's Biggest Business.* Katy, TX: Round Oak Publishing, 2000.

Halman, Martin. *Chemical Fixation of Carbon Dioxide: Methods for Recycling CO_2 into Useful Products.* Boca Raton, FL: CRC Press, 1993. Technical.

Higman, Christopher, and Maarten van der Burgt. *Gasification*. Burlington, MA: Elsevier Science, 2003. Semitechnical.

Hyne, Norman J. *Nontechnical Guide to Petroleum Geology, Exploration, Drilling and Production*. 2nd ed. Tulsa, OK: Penn Well, 2001.

Kemp, William H. *Biodiesel: Basics and Beyond.* Tamworth, ON: Aztext Press, 2006. Everything about biodiesel, including how to make your own.

Larminie, James, and Andrew Dicks. *Fuel Cell Systems Explained.* New York: Wiley, 2000. Semitechnical.

Learsy, Raymond. *Over a Barrel: Breaking the Middle East Oil Cartel.* Nashville, TN: Nelson Current, 2005. An experienced commodities trader exposes how OPEC swindles the world.

Olah, George, Alain Goeppert, and G. K. Surya Prakash. *The Methanol Economy*. Weinheim, Germany: Wiley-VCH, 2006. The best summary discussion of methanol chemistry and the economic possibilities that it enables.

Pahl, Greg. *Biodiesel: Growing a New Energy Economy.* White River Junction, VT: Chelsea Green, 2005.

Roberts, Paul. *The End of Oil: On the Edge of a Perilous New World.* New York: Houghton Mifflin, 2004. Contains a great deal of interesting anecdotal information, but lacks a clear idea of the solution.

Romm, Joseph J. *The Hype about Hydrogen: Fact and Fiction in the Race to Save the Climate.* Washington, DC: Island Press, 2004.

Smil, Vaclav. *Energy at the Crossroads: Global Perspectives and Uncertainties.* Cambridge, MA: MIT Press, 2003. Contains much good data, but the book is defaced by the author's deep Malthusian bias.

Stelter, Stan. *The New Synfuels Pioneers: A History of the Dakota Gasification Company and the Great Plains Synfuels Plant.* Bismarck, ND: Dakota Gasification, 2001. A history of one major project to produce synthetic fuels from coal.

Supp, Emil. *How to Produce Methanol from Coal.* Berlin: Springer-Verlag, 1990. A very good semitechnical treatment.

Tertzakian, Peter. *A Thousand Barrels a Second: The Coming Oil Break Point and the Challenges Facing an Energy Dependent World.* New York: McGraw-Hill, 2006. The view from the perspective of an oilman.

US DOE Energy Information Administration. "International Energy Outlook 2006." Online at http://www.eia.doe.gov/oiaf/ieo/world.html (accessed March 14, 2007).

Yergin, Daniel. *The Prize: The Epic Quest for Oil, Money, and Power.* New York: Touchstone Books, 1991. This book is the single best source for the history of the role of oil in world affairs.

Zacharias, Andrew Paul. *Shuck the Sheiks: Replacing Bloody Middle Eastern Oil with Clean Domestic Ethanol.* Lincoln, NE: iUniverse, 2005.

ON WORLD DEVELOPMENT

Clawson, Patrick, and Rensselaer Lee III. *The Andean Cocaine Industry.* New York: St. Martin's Griffin, 1998. A bit dated, but authoritative and highly readable.

Easterly, William. *The Elusive Quest for Growth: An Economist's Adventures and Misadventures in the Tropics.* Boston, MA: MIT Press, 2002.

———. *The White Man's Burden: Why the West's Efforts to Aid the Rest Have Done So Much Ill and So Little Good.* New York: Penguin, 2006. The best presentation of the conservative viewpoint on the problems of world development. For a good liberal counterpoint, see the works by Stiglitz cited below.

Hayek, F. A. *The Road to Serfdom.* Chicago: University of Chicago Press, 1994. Originally published in 1944, this very powerful book is the Bible of modern libertarian economic and social thought. It is based on its teachings, degraded into dogma, that many self-described libertarians (or classical liberals) oppose "on principle" a government mandate for flex-fuel vehicles. They would do well to read the book again, as Hayek himself is very clear about the need for government action to protect the free market against monopolies. For example, see p. 21: "There is, in particular, all the difference between deliberately creating a system within which competition will work as beneficially as possible and passively accepting institutions as they are. Perhaps nothing has done so much harm to the liberal cause as the wooden insistence of some liberals on certain rough rules of thumb, above all the principle of laissez faire."

Luna, Francisco Vidal, and Herbert S. Klein. *Brazil since 1980.* Cambridge: Cambridge University Press, 2006.

Prahalad, C. K. *The Fortune at the Bottom of the Pyramid: Eradicating Poverty through Profits.* Upper Saddle River, NJ: Wharton School Publishing, 2005.

Sachs, Jeffrey. *The End of Poverty: Economic Possibilities for Our Time*. New York: Penguin, 2005. The UN establishment viewpoint, but still very much worth reading.

Sen, Amartya. *Development as Freedom*. New York: Anchor Books, 2000.

Stiglitz, Joseph. *Globalization and Its Discontents*. New York: W. W. Norton, 2002.

Stiglitz, Joseph, and Andrew Charlton. *Fair Trade for All*. Oxford: Oxford University Press, 2005. A very good source for understanding the problems that caused the failure of the Doha Round Trade Talks.

Yergin, Daniel, and Joseph Stanislaw. *The Commanding Heights: The Battle for the World Economy*. New York: Touchstone Books, 1998. A very readable history of international economic change since World War II.

ON GLOBAL WARMING, THE ENVIRONMENT, AND CARBON-FREE ENERGY

Asmus, Peter. *Reaping the Wind: How Mechanical Wizards, Visionaries, and Profiteers Helped Shape Our Energy Future*. Washington, DC: Island Press, 2001. A journalistic account of the growing wind power industry in the United States.

Bailey, Ronald, ed. *Global Warming and Other Eco-Myths: How the Environmental Movement Uses False Science to Scare Us to Death*. Roseville, CA: Prima Press, 2002. The title says it all.

———. *The True State of the Planet: Ten of the World's Premier Environmental Researchers in a Major Challenge to the Environmental Movement*. New York: Free Press, 1995.

Beckman, Petr. *The Health Hazards of Not Going Nuclear*. Boulder, CO: Golem Press, 1976. This is a delightfully irreverent book that rips the anti-nuclear movement to pieces. A classic.

Bodansky, David. *Nuclear Energy: Principles, Practices, and Prospects*. New York: Springer Science+Business Media, 2004. Semitechnical, but quite well written and very solid.

Gipe, Paul. *Wind Power: Renewable Energy for Home, Farm, and Business*. Rev. ed. White River Junction, VT: Chelsea Green, 2004. The best all-around introduction to wind energy, from one of the leading experts in the field.

Gore, Al. *Earth in the Balance*. New York: Plume Books, 1993. Gore lays out his belief in environment-centered social ethics.

———. *An Inconvenient Truth: The Planetary Emergency of Global Warming and What We Can Do about It.* New York: Rodale Books, 2006. A dire warning of The Wrath to Come from he who would save us.

Hayden, Howard C. *The Solar Fraud: Why Solar Energy Won't Run the World.* 2nd ed. Pueblo West, CO: Vales Lakes Publishers, 2004. Rather strident, but a good antidote to the innumerable solar hype books on the market.

Heaberlin, Scott. *A Case for Nuclear-Generated Electricity (Or Why I Think Nuclear Power Is Cool and Why It Is Important that You Think So Too).* Columbus, OH: Battelle Press, 2003. A very readable and informal, yet technically sound treatment of the subject of nuclear energy and its environmental benefits.

Houghton, John. *Global Warming: The Complete Briefing*. 1994. Reprint, Cambridge: Cambridge University Press, 2004. A detailed scholarly presentation of the case for global warming.

Huber, Peter. *Hard Green: Saving the Environment from the Environmentalists.* New York: Basic Books, 1999.

Lomberg, Bjorn. *The Skeptical Environmentalist.* Cambridge: Cambridge University Press, 2001. This is a very interesting and controversial book. Lomberg is a Danish socialist who decided to make a critical evaluation of environmentalist ideas across the board, finding many instances in which they were detrimental to both social and environmental welfare. Well worth reading.

Michaels, Patrick J. *Meltdown: The Predictable Distortion of Global Warming by Scientists, Politicians, and the Media.* Washington, DC: Cato Institute, 2004. A no-holds-barred assault on the global warming theory.

Morris, Robert C. *The Environmental Case for Nuclear Power: Economic, Medical, and Political Considerations.* New York: Continuum International, 2000.

Philander, S. George. *Is the Temperature Rising? The Uncertain Science of Global Warming*. Princeton, NJ: Princeton University Press, 1998. A measured assessment supporting global warming by a Princeton geoscientist.

Schneider, Stephen, and Randi Londer. *The Coevolution of Climate and Life.* San Francisco, CA: Sierra Club Books, 1984. A good presentation of the science of climate change, written by leading researchers prior to the current controversy.

Sweet, William. *Kicking the Carbon Habit: Global Warming and the Case for Renewable and Nuclear Energy.* New York: Columbia University Press, 2006. An engineer offers a pragmatic approach to reducing carbon dioxide emissions from coal-fired power plants.

ON WORLD WARS I AND II

Churchill, Winston. *The Second World War.* New York: Houghton Mifflin, 1948. A titanic five-volume history of the war by the greatest of the war leaders.

———. *The World Crisis, 1911–1918.* New York: Charles Scribner's Sons, 1931. Reprint, New York: Free Press, 2005.

Eisenhower, John S. D. *The Bitter Woods: The Battle of the Bulge.* New York: G. P. Putnam's Sons, 1969. Well written, but I prefer Toland's earlier and more GI-centered account, cited below.

Ferguson, Niall. *The War of the World: Twentieth-Century Conflict and the Descent of the West.* New York: Penguin, 2006.

Giles, Janice Holt. *The Damned Engineers.* New York: Houghton Mifflin, 1970. This book is a real gem. Janice Holt Giles was the wife of Sergeant Henry Giles, who served with the 291st Engineer Combat Battalion at the Battle of the Bulge. The 291st was the unit that stopped Peiper's 1st SS Panzer division from taking the American fuel depot at Stavelot. Giles tells the story from the GI's perspective.

Grossman, Vasily. *Life and Fate.* Translated by Robert Chandler. New York: Harper & Row, 1985. Grossman was a dissident Soviet Jewish physicist, but during World War II he served as a combat correspondent, covering the battle of Stalingrad for the army newspaper *Red Star.* After the war, he wrote up his experience in the form of this phenomenal Tolstoyan epic novel. There are scenes in this book you will not forget for as long as you live. In the opinion of many, it is the greatest work of twentieth-century Russian literature. The book was banned by the Soviet authorities because, despite its patriotic theme, it presented far too penetrating an analysis of the nature of the totalitarian mind.

Hornfischer, James D. *The Last Stand of the Tin Can Sailors.* New York: Bantam Books, 2004. The latest full account of the battle of Leyte Gulf.

Ito, Masanori. *The End of the Imperial Japanese Navy*. New York: McFadden Books, 1965. The Pacific war as seen by the Japanese navy.

Keegan, John. *The First World War*. New York: Knopf, 1999.

———. *The Second World War*. New York: Penguin, 1989.

Neillands, Robin. *The Bomber War*. New York: Barnes and Noble Books, 2001.

Overy, Richard. *Why the Allies Won*. New York: W. W. Norton, 1995. A very good scientific history critically examining the causes of the Allied victory.

Speer, Albert. *Inside the Third Reich*. New York: Avon Books, 1970. Hitler's minister of industry tells all.

Toland, John. *Battle: The Story of the Bulge*. New York: Random House, 1959.

Werth, Alexander. *Russia at War: 1941–1945*. New York: Avon, 1964. Werth was the only Western journalist to spend the entire war in Russia and survive. His book is incredible.

Woodward, C. Vann. *The Battle for Leyte Gulf*. New York: Ballantine Books, 1946. The classic account of what was both the largest fleet action of the Pacific war and the US Navy's finest hour.

ON FUSION POWER, SPACE EXPLORATION, AND THE HUMAN FUTURE

Fowler, T. Kenneth. *The Fusion Quest*. Baltimore, MD: Johns Hopkins University Press, 1997. A mostly nontechnical account of the fusion effort by a thirty-year veteran program leader.

Gross, Robert A. *Fusion Energy*. New York: Wiley, 1984. A technical book, heavy on the math and theory, but with some nontechnical and semi-technical parts accessible to the more general reader.

Heppenheimer, T. A. *The Man-Made Sun: The Quest for Fusion Power*. New York: Little, Brown, 1984. Dated, but still worth reading.

Herman, Robin. *Fusion: The Search for Endless Energy*. Cambridge: Cambridge University Press, 1990. A lively, nontechnical account of the fusion program by a professional journalist.

Lewis, John, and Ruth Lewis. *Space Resources: Breaking the Bonds of Earth*. New York: Columbia University Press, 1987.

Mendell, Wendell. *Lunar Bases and Space Activities of the 21st Century.* Houston, TX: Lunar and Planetary Institute, 1995.

National Commission on Space. *Pioneering the Space Frontier: An Exciting Vision of Our Next Fifty Years in Space.* New York: Bantam, 1986.

Sagan, Carl. *Pale Blue Dot*. New York: Random House, 1994.

Simon, Julian. *The Ultimate Resource 2*. Princeton, NJ: Princeton University Press, 1996. An update of a 1983 work by Simon skewering Malthusianism. Shortly after the earlier version was published, Simon made a set of bets with *Population Bomb* (New York: Ballantine Books, 1968) author Paul Ehrlich comparing their predictions on the future price of an array of resources. Simon won all of them.

Simon, Julian, and Herman Kahn. *The Resourceful Earth: A Response to "Global 2000."* New York: Basil Blackwell, 1984. A refutation of the Malthusian Club of Rome.

Stacey, Weston. *Fusion: An Introduction to the Physics and Technology of Magnetic Confinement Fusion*. New York: Wiley, 1984. Semitechnical.

Zubrin, Robert. *Entering Space: Creating a Spacefaring Civilization.* New York: Tarcher Putnam, 1999. Humanity's space prospect, from low Earth orbit to the stars.

Zubrin, Robert, with Richard Wagner. *The Case for Mars: The Plan to Settle the Red Planet and Why We Must*. New York: Free Press, 1996. This is my best work on the human future in space. If you read just one book on this subject, this one should be it.

INDEX

ABOUT THE AUTHOR

Robert Zubrin is president of Pioneer Astronautics, an aerospace engineering R&D firm, and also leads the Mars Society, an international organization dedicated to furthering space exploration. For many years he worked as a senior engineer for Lookheed Martin. He holds a doctorate in nuclear engineering, and has nine US patents granted or pending. In addition, he is the author of the critically acclaimed nonfiction books *The Case for Mars*, *Entering Space*, and *Mars on Earth*; the science fiction novels *The Holy Land* and *First Landing*; and articles in *Scientific American*, *New Atlantis*, *American Enterprise*, the *New York Times*, and the *Washington Post*. He has appeared on major media including CNN, C-SPAN, the BBC, the Discovery Channel, the Science Channel, NBC, ABC, and NPR.